Hey By George! III
Uncut Stones

By George W Denn

HEY By George!III
by George W Denn

Printed in the United States of America

ISBN 9781619966178

www.xulonpress.com

Exodus: 24-25 in every place where I record My name I will come to you, and I will bless you. And if you make me an altar of stone, you shall not build it of hewn stone; for if you use your tool on it, you have profaned it.

Foreword

As a hay farmer, one of the tasks I face each year in the spring, is the seemingly un endless task of picking up stones, and rocks that have been deposited on the surface of the soil after cultivating and planting of a field that will be used for hay. Although this practice has no monetary value in itself, it is very necessary to prevent costly break downs, and sometimes total destruction to a piece of haying equipment which can cost thousands of dollars. By learning this lesson the hard way I now try to pick up anything bigger than a half dollar! These stones are then deposited in piles near the field some being quite large after years of adding to them. In these Piles you will find a variety of stones ranging from very small to huge. So large sometimes that you need the help of others, and larger equipment to get them removed! Hey By George! III Uncut Stones. Is a collection of stories that I have written over 3 years' time in addition to Hey By George! And Hey By George! II on the Northwest side of Wita Lake. In them you will read of various things that I have had to deal with, and how I dealt with them on my spiritual journey. Like stones If things aren't dealt with that the Lord is bringing your way they may come back to haunt you! Like the subtitle Uncut stones the stories that you are about to read, have been unedited, beyond my hand so just to warn you there may be some wrong punctuation, sentence structure, and misspelled words! For this I apologize and hope it won't diminish the message they contain! I hope you enjoy reading these 15 stories, and hope you can relate to them from stories that have happened in your own life, and so I add to the other stories I have written these 15 to make my monument of Uncut Stones!

~George W Denn

Acts 16:16-31: *Once when we were going to the place of prayer, we were met by a slave girl who had a spirit by which she predicted the future. She earned a great deal of money for her owners by fortune-telling. This girl followed Paul and the rest of us, shouting, "These men are servants of the most high God, who are telling the way to be saved," She Kept this up for many days. Finally Paul became so troubled that he turned around and said to the spirit, "In the name of Jesus Christ I command you to come out of her!" At that moment the spirit left her. When the owners of the slave girl realized that there hope of making money was gone, they seized Paul and Silas and dragged them into the marketplace to face the authorities. They brought them before the magistrates and said, "These men are Jews, and are throwing our city into an uproar by advocating customs unlawful for us Romans to accept or practice." The crowd joined in the attack against Paul and Silas, and the magistrates ordered them to be stripped and beaten. After they had been severely flogged, they were thrown into prison, and the jailer was commanded to guard them carefully. Upon receiving such orders, he put them in the inner cell and fastened there feet in the stocks. About midnight Paul and Silas were praying and singing hymns to God, and the other prisoners were listening to them. Suddenly there was such a violent earthquake that the foundations of the prison were shaken. At once all the prison doors flew open, and everybody's chains came loose. The jailer woke up, and when he saw the prison doors open, he drew his sword and was about to kill himself because he thought the prisoners had escaped. But Paul shouted, "Don't harm yourself! We are all here!" The jailer called for the lights, rushed in and fell trembling before Paul and Silas. He then brought them out and asked, "Sirs, what must I do to be saved?" They replied, "Believe in the Lord Jesus, and you will be saved- you and your household."*

Well it finally happened. Today I had to go to prison! Like other followers of Jesus Christ I finally ended up in a place I never wanted to be! But I was only there for about an hour then they let me go! I was messing around on my computer the other day. When around 3 in the afternoon my cell phone started ringing. I noticed it to be a strange area code so I didn't answer it. I figured they would leave a message if it was important. They did leave a message, and while I was retrieving the Message they called again. I didn't want to answer it until I listened to their message. The message was from the Minnesota treatment center from a guy named Rick. I thought to myself who in the world do I know by that name that could be there! I thought it may be someone I know that's having a problem with alcohol or something. But later found out that's a prison were they keep sexual offenders! It's kind of ironic that St. Peter who was once put in prison now has a town named after him with a prison in it!

> ***Acts 12: 5-10:*** *So Peter was kept in prison, but the church was earnestly praying to God for him. The night before Herod was to bring him to trial, Peter was sleeping between two soldiers, bound with two chains, and sentries stood guard at the entrance. Suddenly an angel of the Lord appeared and a light shown in the cell. He struck Peter on the side and woke him up. "Quick, get up!" he said, and the chains fell off Peters wrists. Then the angel said to him, "Put on your clothes and sandals." And Peter did so. "Wrap your cloak around you and follow me," the angel told him. Peter followed him out of the prison, but had no idea that what the angel was doing was really happening; he thought he was seeing a vision. They passed the first and second Guards and came to the Iron Gate leading to the city. It opened for them by itself, and they went through it. When they had walked the length of one street, suddenly the angel left him.*

Well I started to feel bad. I thought maybe someone I knew needed some help. So I said a short prayer, and asked God if this was something that needed my attention; to please have this person call one more time. About a half hour latter another phone call came from a man that said he worked at the St. Peter regional treatment center. He was telling me that this Rick guy was trying to call me about a load of wood but I refused his call! I told the man that I hadn't necessarily refused his call; I just hadn't answered the phone. I was asking this guy if this Rick was in prison how was he going to receive and pay for the load. The man informed me that Rick was a Native American. And the load of wood was for a sweat lodge, something to do with an Indian religious ceremony. Oh just what I need I thought; here I am a Christian and I am on my way to promoting some pagan thing! Then I thought I like Rick am too in prison, but I am a prisoner of Christ and my life belongs to him.

> ***Ephesians 4:1:****As a prisoner for the Lord, than, I urge to live a life worthy of the calling you have received.*

So with all this in mind when my phone rang again, and I saw the number; I quickly answered and followed the instructions! The next thing you know I was talking with This Rick. He asked all about the wood, and was also wondering if I could get him about 50 field stones, between the size of a grape fruit and a small melon? I told him I had some pretty fair wood but the stones would have to wait until the snow melts in the spring time. Rick was telling me it was pretty hard to communicate with people outside his facility with all the rules and regulations they had. I was somewhat amazed that he could talk to anyone at all! He told me another lady would be in contact with me to confirm all this and arrange payment for the wood. She called and confirmed all this and said payment would take about a week because it has to come from Moose Lake, but she was wondering if I could deliver it now and wait for payment? "Sure," I said, "I do that for others so I will do it for you I want to be Helpful!" Also if they liked the wood they would continue to buy it from me, because it is so hard to get people to bring wood in to them. I may be wrong but I have the feeling the money will be coming from the gambling casino up there in Moose lake! Ok now we have a pagan ritual, sexual immorality, and gambling! I thought what next, but God quickly shows me, I too have been guilty of sinful things in the past, but since Jesus has cleansed me of these things I am no longer guilty!

> ***Romans 8:1-2:*** *Therefore, there is now no condemnation for those who are in Christ Jesus, because through Christ Jesus the law of the Spirit of life set me free from the law of sin and death.*

I also know God reaches us all where we are at even if that's in prison

> ***Isaiah 49:8-9:*** *This is what the Lord says: " In the time of my favor I will answer you, and in the day of salvation I will help you; I will keep you and make a covenant for the people, to restore the land and to reassign its desolate inheritances, to say to the captives, 'Come out,' and to those in darkness, 'Be free!' "They will feed beside the roads and find pasture on every barren hill.*

In some ways this all gives me the creeps! So what do I do first? I called my friend Jeff and told him the story and I thought we should pray over this load, because I felt God may be up to something here. And with all these other spirits involved. I wanted to make sure we invoked the Father, Son, and Holy Spirit the one that's above all other spirits and not take any chances! And that's what we did after I loaded the load of wood. It's interesting that I've notice in my walk with the Lord a lot of times he starts something with a load of wood! Also when Jeff and I are involved with what the Lord is up to the time 12:22 PM is generally the start or finish of something! Guess when Jeff walked into my house I showed him the time on the stove 12:22 PM! He laughed and said he never really knew what time it was. We prayed over the load of wood and asked God to bless it, the blood of Jesus Christ to cover it, and the Holy Spirit to be in and around every piece of would and that his purposes for this load of wood be fulfilled. Now there may be some that feel what we had done was unnecessary but I for one wasn't going to take any chances, besides it wasn't going to hurt anything either!

> ***Matthew 12:25-30:*** *Jesus knew their thoughts and said to them, "Every kingdom divided against itself will be ruined, and every city or household divided against itself will not stand, If Satan drives out Satan, he is divided against himself. How than can his kingdom stand? And if I drive out demons by Beelzebub, by whom do your people drive them out? So than, they will be your judges. But if I drive out demons by the Spirit of God, then the kingdom of God has come upon you. "Or again, how can anyone enter a strong man's house and carry off his possessions unless he first ties up the strong man? Then he can rob his house. "He who is not with me is against me, and he who does not gather with me scatters.*

I had to be at the guard house at 1:30 with my load, because at 2:00 o'clock they change the guard and that would not be a good time to show up according to the religious coordinator. I had to show the guard my driver's license so I could prove I was who I said I was! They gave me pass #29 with a map on back and showed me where I needed to go, and raised the arm so I could proceed into the prison. I only got lost twice! This place was originally built as an asylum for the insane. I drove past the old square limestone asylum. Its steep stair steps were barricaded so no one could enter there. I heard of this place often when I was a kid. My mother called it the St. Peter nuthouse! A place that she was sure to end up at; if us kids didn't start to mind her better! As I drove by it I had visions of people inside grabbing the bars in front of the windows, and screaming for all they were worth! But then again, I have always had a great imagination! I had to sit and wait about 20 minutes before someone showed up to let me in the gate. As I waited I was looking at the fence. The razor wire coiled all across the top, and all along the sides. I was thinking I wish I would have had a fence like this when I was still raising livestock! They would have never gotten lose from there! Then I was trying to figure out how I would climb it if I were one of those on the other side. Finally the gate slowly opened and I was waved into an in between section where the guards could search my loaded vehicle for things like drugs, fire arms, tobacco, or alcohol. They had me open the hood to look under it. They said I was clean, but little did they know that the Holy Spirit was lurking in every piece of wood in the box; not to mention Jesus Living inside of me! They gave me a club to lock my steering wheel and I was told to keep the key for it along with my ignition key on my person at all times. The prison door slowly opened and I was told to drive straight ahead to the next guard I saw for further instructions! I was starting to feel like I was part of an old Hogan's hero's movie! The guard rode with me; he showed me a side walk I was supposed to drive down. I asked if I happened to get stuck do you have something available to pull me out with. Of course not! We drove right up in front of a frame for a wigwam that had three plastic chairs in it, and one was tipped over. I was told to stop. And was told that three prisoners would come and unload the wood, which didn't bother me any! I asked the guard if I should lock the doors. "Not with me here," he said. But I held the keys for the club and the ignition as I was told earlier. As the prisoners unloaded the wood; I asked the guard if this frame of a wigwam was the sweat lodge. "Yes," he said. "It's supposed to have buffalo hides on it, but you can't have everything, they cover it with a canvas tarp," he mentioned. I asked if this was considered a minimum or a maximum security prison. "Pretty much maximum for the sexual offenders," he said. "Are the wires electrified, "I asked. "Only the ones on the sides are motion sensitive the ones across the top are not," he said. "I see," I said. Just then one of the guys slipped on the snow and fell under the pickup. The guard was telling me it was a good thing that wasn't serious or he would be here until 4:00 o'clock filling out paper work! "You wouldn't believe the amount of rules and regulations we have to follow around here, "He mentioned! He also said something to the effect that the state had about a 5 billion dollar deficit, and he was wondering if that would have any effect on the facility? I just told him I knew all about deficits, and it was my experience that they were not too good! All of a sudden my phone rang a man was asking about hay and I eventually sold him $4500 worth of hay something I felt a little strange as I hadn't sold much to speak of since last October! This is another sign that

God may be up to something here; as I am usually messing around with hay when I see him at work! Just then the religious coordinator walked up she was asking the prisoners how they liked the wood. It was some real good stuff they told her! She was also asking me about the stones. I told her that I could get them some come spring time. She mentioned it sounded like I would probably be back a couple of times before then! After I was unloaded I mentioned to the guard that I would like my gloves back. He had given them to one of the guys, and they looked like they had forgotten about it! I backed up, and drove to the gate. The guard there told me to get out of the vehicle and walk around it. I was thinking I sure am at the mercy of the guards here, and I sure will be glad to be on the other side of those gates! The first gate opened and I drove through. I stopped and opened the hood so he could check under it, and I gave him the club back. Ok the guard said you are free to go. I got back in the pickup and the gate opened up, and I was free! I was thinking how much better it is to belong to Jesus Christ.

> ***John 10 :1-16:****"I tell you the truth, the man who does not enter the sheep pen by the gate, but climbs in by some other way, is a thief and a robber. The man who enters by the gate is the shepherd of his sheep. The watchman opens the gate for him, and the sheep listen to his voice. He calls his own sheep by name and leads them out. When he has brought out all his own, he goes on ahead of them, and his sheep follow him because they know his voice. But they will never follow a stranger; in fact, they will run away from him because they do not recognize a stranger's voice." Jesus used this figure of speech, but they did not understand what he was telling them. Therefore Jesus said again, "I tell you the truth, I am the gate for the sheep, All who ever came before me were thieves and robbers, but the sheep did not listen to them. I am the gate; whoever enters through me will be saved. He will come in and go out, and find pasture. The thief comes only to steal and kill and destroy; I have come that they may have life, and have it to the full. "I am the good shepherd. The good shepherd lays down his life for the sheep. The hired*

hand is not the shepherd who owns the sheep. So when he sees the wolf coming, he abandons the sheep and runs away. Then the wolf attacks the flock and scatters it. The man runs away because he is a hired hand and cares nothing for the sheep. "I am the good shepherd; I know my sheep and my sheep know me- just as the Father knows me and I know the Father- and I lay down my life for the sheep. I have other sheep that are not of this sheep pen. I must bring them also. They too will listen to my voice, and there shall be one flock and one shepherd.

Then to belong to this place with its rules and regulations and pagan rituals and guards everywhere. What is God up to with all of this? Who knows!

John 3:8: *The wind blows where it pleases. You hear its sound, but you cannot tell where it comes from or where it is going. So it is with everyone born of the Spirit."*

It is my prayer as the Indians do there sweat lodge religious thing. That they will see our Triune God in the fire, and he will speak to them just as he did with Moses, in the burning bush!

Exodus 3:1-5: *Now Moses was tending the flock of Jethro his father-in -law, the priest of Midian, and he lead the flock to the far side of the desert and came to Horeb, the mountain of God. There the angel of the Lord appeared to him in flames of fire from within a bush. Moses saw that though the bush was on fire it did not burn up. So Moses thought," I will go over and see this strange sight-why the bush does not burn up." When the Lord saw that he had gone over to look, God called to him from within the bush, "Moses! Moses!" And Moses said, "Here I am."*

"Do not come any closer," God said. "Take off your sandals, for the place where you are standing is holy ground."

I drove back past the creepy old asylum down to the guard house.

The guard said, "I see you made it back!"

"Yep," I said. As I handed her back the pass. As I drove back to the freeway I had to laugh, because it's kind of ironic. My mother never did end up here like she used to tell us she would, but I may find my way back to this prison. God willing! God does have the greatest sense of humor, doesn't he! So just remember this,

> ***John 8:34-37:*** *Jesus replied, "I tell you the truth, everyone who sins is a slave to sin. Now a slave has no permanent place in the family, but a son belongs to it forever. So if the Son sets you free, you will be free indeed.*

Gods' peace and abundant blessings to you all!

Your Brother in Christ,

George Denn

Hey By George! February 4 2009

> ***Psalm 23:1-3:*** *The Lord is my shepherd; I shall not be in want. He makes me lie down in green pastures, he leads me beside quiet waters, he restores my soul. He guides me in paths of righteousness for his name's sake.*

It seems to me that the only time I hear psalms 23 read is at funerals! I don't think God had it put in his word, nor do I think King David wrote it for that purpose only. So here in the dead of a Minnesota winter I will try and use it for something that pertains to life! For years now I have complained how I hate winter. I used to have to work outside in it all day long, and it seemed to me that I would end up wrecking more stuff, and costing me more money to fix what I would wreck, than the money I would make. So I came up with a solution. I decided I would do my work in spring, summer and fall and do as little as possible during the winter! So far this works out well for me; it has caused some people I know to wonder about me so I figure I will just let them wonder! Now that I have things arranged in my life that I don't have much to do in the winter. I Find that now I dislike it because there is nothing much to do! Most of my winter days are spent reading books reading the bible and writing these stories. Something that I feel is not a waste of time. But yet I still dislike winter! One day I thought winter is a season that was created by God, and anything that God creates has to be good! So one day I asked God, "God can you please show me something about winter time that I can like!" God pointed out to me that it is a time for rest and rejuvenation. "Hmm" I said I never really thought of that! So while the Lord is restoring my soul from last year, and preparing me for this one. I have been thinking about money, and why we work for it?

Some of us humans work for money, because we love money, and have no respect for God.

Some of us humans work for money, because we love money, but we try to serve God also.

Some of us humans work for God, and make no money.

Some of us humans work for God, and make lots of money, but hardly give money a thought.

My theory about money is one most of you have heard me say before. "Money is like manure it is no good unless you spread it around." And yes I did borrow that phrase from the movie Hello Dolly!

Time and time again the bible tells us the right way and the wrong way to look at money. I heard a guy say once that money has God like powers because it is here before you are, and here after you leave! It will also do things for you, and you can see, feel and touch it.

God will do way more for us and has, than money ever could, and yes you can see feel and touch God however you must re lie on his Holy Spirit to do so! God also puts much more value in what he has to offer than the things that we value! Time after time in God's word it is mentioned how important the things we hold important really are.

> ***Isaiah 55:1-3:*** *"Come, All you who are thirsty, come to the waters; and you who have no money, come, buy and eat! Come; buy wine and milk without money and without cost. Why spend money on what is not bread, and your labor on what does not satisfy? Listen, listen to me, and eat what is good, and your soul will delight in the richest of fare. Give ear and come to me; hear me that your soul may live. I will make an everlasting covenant with you, my faithful love promised to David.*

Jesus himself tells us how we should use the money we have!

> ***Luke 16:9:*** *I tell you use worldly wealth to gain friends for yourselves, so that when it is gone, you will be welcomed into eternal dwellings.*

If Jesus says someday all our money will be gone; it will all someday be gone! Also notice what Jesus says in,

> ***Luke 16:13:*** *"No servant can serve 2 masters. Either he will hate the one and love the other, or he will be devoted to one and despise the other. You cannot serve both God and Money."*

Jesus also mentions in,

> ***Luke 12:15:*** *Then he said to them, "Watch out! Be on your guard against all kinds of greed; a man's life does not consist in the abundance of his possessions."*

Thinking of these scriptures reminds me of something God taught me a few years ago during a trip to Oklahoma at Thanksgiving time. How money can and often very easily does, dethrone God and what he is trying to do in our lives!

About 3 years ago at Thanksgiving time I was invited down to Oklahoma to spend Thanksgiving Day with some friends down there. It's about a nine hour drive from here to there, so this time I decided to fly down there. It often amuses me how whenever I fly south I have to drive an hour in the wrong direction to get to the airport. David Claggett made my flight arrangements so I had a $240 check to give him when I got down there to pay for my flight. I decided to stretch myself and not take any luggage with me. All I took was my cell phone and wallet and the clothes on my back. Jeff drove me to the airport in Minneapolis and Brad picked me up in Tulsa a few hours later. The first thing we did was go to Wal-Mart's and buy a couple pair of clothes for me to wear while I was there. I was really paying attention to what was going on also, because I was on sermon duty the following week at church. So I was looking for some direction from God for a message! The annual Claggett Reunion is held at the First Baptist church there in Vinita Oklahoma. Every Thanksgiving day for three years I was a guest. After a huge Dinner this year they had some prizes. I won a cap for traveling the farthest distance to get there. It was a white cap promoting Edward D. Jones investment co. My old cap was getting slightly worn it promoted Promise Keepers an organization promoting Gods values in the lives of men. I no sooner placed the new cap on my head. When God spoke to me just how easily the things of money try to Dethrone Him and the purposes he has for our lives. I quickly put my old cap back on and decided I would wear the new on here on the farm!

> ***1 Kings 19:11-13:*** *The Lord said, "Go out and stand on the mountain in the presence of the Lord, for the Lord is about to pass by." Then a great and powerful wind tore the mountains apart and shattered the rocks before the Lord, but the Lord was not in the wind. After the wind was an earthquake, but the Lord was not in the earthquake. After the earthquake came a fire, but the Lord was not in the fire, And after the fire came a gentle whisper. When Elijah heard it, he pulled his cloak over his face and went out and stood at the mouth of the cave. Then a voice said to him, "What are you doing here, Elijah?"*

What was I doing in Oklahoma? Besides visiting friends I was trying to get a message from the Lord for my sermon! Before we left I tried to Give David the check for my flight, but he wouldn't take it. He said for all that I had done for his son there was no way he would accept it! Brad, Kelly and I left for Oklahoma City after the reunion. Since it looked like I had no need for money on this trip I decided to give Brad $100 for gas for hauling me around! I mean you have to spend something on a trip don't you? When we got to Oklahoma City I met Brad's uncle Poe. He seemed quite amused by my stories of my pumpkin farm. Like most people he seemed to think I was making tons of money raising pumpkins. What most people don't realize is that I raise pumpkins in my service to the Lord. I don't raise them to make money. Yes I do make a living raising pumpkins but doesn't God promise that any way?

> ***Matthew 6:25-34:*** *"Therefore I tell you, do not worry about your life, what you will eat or drink; or about your body what you will wear. Is not life more important than food, and*

the body more important than clothes? Look at the birds of the air; they do not sow or reap or store away in barns, and yet your heavenly Father feeds them. Are you not much more valuable than they? Who of you by worrying can add a single hour to his life? "And why do you worry about clothes? See how the lilies of the field grow. They do not labor or spin. Yet I tell you that not even Solomon in all his splendor was dressed like one of these. If that is how God clothes the grass of the field, which is here today and tomorrow is thrown into the fire, will he not much more clothe you, O you of little faith? So do not worry, saying, 'What shall we eat?' or 'What shall we drink?' or 'What shall we wear?' For the pagans run after all of these things and your heavenly Father knows that you need them. But seek first his kingdom and his righteousness, and all these things will be given to you as well. Therefore do not worry about tomorrow, for tomorrow will worry about itself. Each day has enough trouble of its own.

The next day Larry, Brads mothers husband; took me and Poe out to Poe's farm near Gary Oklahoma. Poe was a gunner and an engineer on the boomer Plane's in WWII. After the war Poe came back to Oklahoma and once again took up farming. One of Poe's greatest projects during his life was to build a house/ grain elevator/ barn All rolled into one building. It was designed to generate its own power; it is a colossal structure! The best way to describe it is what the local people have named it, "The continuance"! They call it that, because Poe never really finished it. Poe's age crept up on him like it does most of us, and he was unable to finish his project! Time and the elements are eroding the structure away. Poe told me that he had a bunch of black powder and nitro glycerin stored in there yet. If someone lights a match to clean the place up I do believe that Gary Oklahoma will remember Poe with a bang! I told Brad about this so hopefully no one gets killed! After we left Poe's farm we ventured to the Grave site of Jesse Chisholm; he was the man who created the Chisholm Trail. Part of this trail still is able to be seen on the edge of Poe's farm. If you look him up in the encyclopedia you can get some idea on what he all did! Jesse Chisholm did a lot of great things for this country, but his life was largely about serving money, and he died at age 62 when he ate a bad piece of bear meat that one of his Indian friends had cooked in a cooper kettle! Now some 150 years after he has lived this once important man is barely remembered by most! By the looks of things around his grave site you can hardly see any activity that people come out here to visit his grave. On the way back to Oklahoma City that afternoon I was thinking about Poe's life and Jesse Chisholm's life, and then I started thinking about the apostle Paul's life and how he served God after his conversion experience on his way to Damascus.

> ***Acts 9:1-9:*** *Meanwhile, Saul was breathing out murderous threats against the Lord's disciples. He went to the high priest and asked him for letters to the synagogues in Damascus, so that if he found any there who belonged to the way, whether men or women, he might take them as prisoners to Jerusalem. As he neared Damascus on his journey, suddenly a light from heaven flashed around him. He fell to the ground and heard a voice say to him, "Saul, Saul, why do you persecute me?"*
>
> *"Who are you Lord?" Saul asked. "I am Jesus, Whom you are persecuting," he replied. "Now get up and go into the city, and you will be told what you must do." The men traveling with Saul stood there speechless; they heard the sound but did not see anyone. Saul got up from the ground, but when he opened his eyes he could see nothing. So they led him by the hand into Damascus. For three days he was blind, and did not eat or drink anything.*

Most of us aren't a whole lot different than Paul running as fast as we can in the opposite direction than the Lord intends for us to go! Most of the time the Lord has to intervene in our lives too; to get our attention! In the same tenacity after his conversion, Paul serves the Lord with all his heart soul and mind. Only once does the bible casually mention that Paul was a tentmaker by trade; and says absolutely nothing about his yearly income or what he has saved for retirement!

> *Acts 18:1-3: After this, Paul left Athens and went to Corinth. There he met a Jew named Aquila, a native of Pontus, who had recently come from Italy with his wife Priscilla, because Claudius had ordered all the Jews to leave Rome. Paul went to see them, and because he was a tentmaker as they were, he stayed and worked with them.*

But yet here we are almost 2000 years after Paul's time, and how he lived and what he wrote is still having a profound impact on us today! Most people if you asked them would be able to tell you who the apostle Paul is. Paul also had a bit to say about money.

> ***1 Timothy 6:3-10:*** *If anyone teaches false doctrines and does not agree to the sound instruction of our Lord Jesus Christ and to godly teaching, he is conceited and under stands nothing. He has an unhealthy interest in controversies and quarrels about words that result in envy, strife, malicious talk, evil suspicions and constant friction between men of corrupt mind, who have been robbed of the truth and who think that godliness is a means to financial gain. But godliness with contentment is great gain. For we bought nothing into the world, and we can take nothing out of it. But if we have food and clothing, we will be content with that. People who want to get rich fall into temptation and a trap and into many foolish and harmful desires that plunge men into ruin and destruction. For the love of money is a root of all kinds of evil. Some people eager for money, have wandered from the faith and pierced themselves with many grief's.*

After a few Days in Oklahoma it was time to fly back home. When Brad dropped me off at Tulsa airport I gave him the clothes I had bought when I first went down there. Brad and Kelly would be coming back to Minnesota sometime I told them to bring them with when they came. I could see how God had orchestrated my visit to give me the sermon material I had needed! Jeff picked me up In

Minneapolis around mid-night and drove me home.

So it looks to me that to invest your money, time and talents. Into Gods kingdom, and to follow Jesus for all you are worth is the best investment any one can make! And when everything is said and done you'll have a portfolio that is out of this world; guaranteed by Jesus Himself!

> ***Mark 10:29-30:*** *"I tell you the truth," Jesus replied, "no one who has left home or brothers or sisters or mother or father or children or fields for me and the gospel will fail to receive a hundred times as much in this present age(homes, brothers, sisters, mothers, children and fields and with them persecutions) and in the age to come, eternal life.*

WOW! What a retirement package! Gods' peace and abundant blessings to you all!

Your brother in Christ,

George Denn

Hey By George! March 25 2009

Proverbs 3:5-6: *Trust in the Lord with all your heart and lean not on your own understanding; in all your ways acknowledge him, and he will make your paths straight.*

Once again winter has given way to spring; my favorite season! Some years this process seems to happen ever so slowly and other years it seems like a violent game of tug of war, between winter and spring. This year it looks like it's going to be the tug of war game! March started out fairly mild, but by the 12th of March it had snowed and was once again winter! We had below zero temps. My thoughts run to the years that I had already had some field work and seeding done by this time; but they are rare maybe twice as I near my 47th year! One always hopes though! I was telling my friend Jeff and Ethan that evening that I felt that the lake east of my house would open up a week early this year. Probably around the 20th. They just looked at one another and smirked and almost simultaneously said "I don't know" in a rather unbelieving way! I said this because the lake had frozen up a week earlier than usual last November. Almost always it freezes the last week in November and goes out the last week in March. This is something I have noticed the years I have spent here. Well the calendar gave me grace the ice went out on the 23rd so this falls on the second to the last week in March. I called Ethan and asked if he noticed if the ice had gone off? His folks live across the lake from me I can just barely see their place from where I sit typing this! Ethan did not dispense me grace though he told me I said the 20th! What are a few days any ways? The fields have been pretty much exposed all winter; so I had it in my mind that quite possibly I could get my wheat seeded on the weekend, or the first part of next week. So when my friend Troy called and asked if I would go to Georgia with him to pick some carpeting up on the weekend; I told him no and revealed to him my wheat planting plans! Troy said he would call me back on Wednesday just in case I had changed my mind. Why is it I can sit around here all winter without any thing to do, and the moment I start even thinking about farm work someone calls me with something to do or there is a meeting I should go too or something, and these events never take place on a rainy day it always seems like it's a nice and sunny day that I can get some work done around here! Or should be getting some work done around here! I could probably be gone somewhere every week from springtime until winter sets in again, but then no farming would get done, and I kind of like to eat, so being that's how I make my living I have to be selective with what I do during that time period. On Wednesday I called Troy and told him I would go with him to Georgia. The fields were not drying as fast as I thought they might, and I didn't want to be sitting around here letting the weather aggravate me! Maybe next week when I get home it would be dry enough to plant wheat? Troy told me to be at his place around 3 PM on Thursday we left at 6 PM. Troy told me that he was sorry that his day hadn't gone as planned. I told Troy to relax because I was used to that! I do believe I have learned the secret to being flexible so that no matter what happens I am not disappointed, but believe me this has not come easy! And it is way better than being disappointed all the time!

> ***Philippians 4:10-13:*** *I rejoice greatly in the Lord that at last you have renewed your concern for me. Indeed, you have been concerned, but you had no opportunity to show it. I am not saying this because I am in need, for I have learned to be content whatever the circumstances. I know what it is to be in need, and I know what it is to have plenty. I have learned the secret of being content in any and every situation, whether well fed or hungry, whether living in plenty or in want. I can do everything through him who gives me strength.*

Anyone who has read about the apostle Paul would probably agree that this secret he learned didn't come easy for him either!

This time our trip to Dalton Georgia was an uneventful one and around 9 am Eastern Time we arrived at our destination. First we had to stop and see Tony one of Troy's Brother in Laws. Tony works at one of the many carpet mills in town; Dalton is the carpet capitol of the world! Tony took us to where Troy needed to pick up some carpet padding. My friend Bob called at this time to see what I was up to. I told Bob I was in Dalton Georgia soaking in the sun, as it was a balmy 60 degrees! Bob seemed somewhat impressed, but said it was only 40 degrees back home! After this Tony treated us to lunch, and after lunch we dropped Tony off so he could finish his day of work. Troy and I drove East to a town called Chatsworth. We were to stay at the home of John and Linda Duckett's; Troy's in laws. I was sitting on a chair on the front porch when their dog came up to me to be petted. John said would you look at that dog he is even nice to Yankees! I just laughed, because I knew John meant it in a teasing way! That afternoon Troy and I drove out to see Mike another of Troy's Brother in laws. I have really enjoyed meeting all these folks from the south. Although they Talk somewhat different then we do up here in the north we all understood one another just fine. I enjoyed our visit with Mike. He lives in the country in a house he built himself on some acreage not too far from where his parent's farm is. I always like talking with Mike he chooses his word carefully and there is a hint of humor when he talks. Mike is of Baptist origin in his faith, and although Mike is going through some times of trouble right now I can't help but feel God is calling Mike to a Closer walk with Thee! Bruce Called while we were talking with Mike he asked us if we were sitting on the ocean beach we told him no but we were next to a swimming pool, but none of us went in yet! Troy, Mike, and I were laughing because the pool we were setting by was still covered for the winter! When we got back to John and Linda's they did a cook out and we all ate on the back porch. Shortly after we ate I went to sleep, and didn't wake up until 7:30 the next morning. After breakfast Tony came over He had found a warehouse that had some carpeting that Troy was looking for. After we had been at the warehouse for some time; Troy asked Tony how much inventory he figured they had there. 2 million dollars' worth was Tony's estimate. I had asked if the economy was affecting the carpet industry. Tony said that it had in a big way. They had 12% unemployment in that town. We had just seen a place the day before where some homeless people were living in tents! In this warehouse they had some hand woven rugs from Pakistan hanging up. The salesman told us that it took two people about 11 months to weave one rug!

I looked at the price tag $4400. Something tells me that the two people that wove the rug didn't get much for their efforts! So I guess farming isn't the only occupation that you don't make much money at! I guess it's just too bad that I don't have a whole lot of interest in houses and stuff like rugs to furnish them. Tony figures that if I had a wife I would probably have a lot more interest! Who knows? That afternoon Mike came by to take Troy and I for a ride up to grassy mountain. Grassy Mountain is on the very south western side of the Appalachian mountain range. We drove up a mountain road that was built in the 1930's by the CCC camps. I was wondering how on earth people got around on this mountain hauling logs and farming before this road was built? I could see why moon shiners would like this country! Mike was telling how his grandmother or great grandmother was a Cherokee Indian and she had a still; she even had a picture of it! Mike didn't know where the picture was, but it was somewhere he said. I was telling Mike that I had some Sioux Indian blood in me from my mother's side. Also I was telling Mike that I think there was some people in my family that had crossed paths with the bootleg whisky trade but it didn't come from my mother's side! We passed the remains of John's grandfather's cabin; all that remains is the stone fireplace and chimney. Later that evening John told us the story when he was a boy there was a big old long table in that cabin. One evening his dad and uncle shot a wild boar, and after they had dress it out brought it home and placed it on the table until the next day. That evening John had gotten up to go to the toilet and he saw that hog lying on the table and it scared the daylights out of him! The image I have in my mind of that just makes me chuckle! As we drove up the mountain road Mike was showing us a draw with a spring running through it. He was catching "spring lizards" I guess they are good to catch fish with! He found an old moon shine still back in that draw that it's owner had long since abandoned. He said there were some jugs of liquor still there, but he never tried any! I remembered a time I had tried some; it is some pretty rough stuff as I recall! Mike took us up to his dad's cabin on the top of the mountain. I pointed to an arrow that I saw sticking out of the ground; I was asking Mike what tribe of Indians he reckoned it was from? "Looks like the Hackeysaw tribe!" Mike said, and we all laughed! Mike took us out behind the cabin and he showed us a trap they use for black bear, and wild hogs. After a brief time inside the cabin we continued on down the mountain a different way than the way we came. Mike showed us a small lake that they often fish on, and after that we past a couple young fellows who were sawing a log with a crosscut saw! Mike was saying that sure looked interesting, and was asking if we wanted to stop and watch; laughing we said we better not! I still have my grandpa Denn's cross cut saw out in the shed. I bought it on his auction when I was 14, but one of the handles is missing; probably the one on my side of the saw! Then we drove back into Chatsworth and our mountain ride was done. We had supper with John and Linda, and our visit was coming to a close. We left the next morning about 4:00 am to come back to Minnesota. I know I sure did enjoy the visit to Georgia. I enjoyed getting to know the folks I meet a year ago a little better. I think Troy and I even got Mike thinking about raising pumpkins this year! It is a good thing I went with Troy because in the last couple days I think we received 2 inches of rain, and we had a little snow this morning! It sounds like it is going to be cold for a few days so it looks like I can put wheat planting on hold for a while!

A couple of days ago I stopped at the implement dealer to get a part for my tractor. I had to talk to Jim he is part owner and longtime friend. We always seem to talk about many things. He was asking me what I thought about the year ahead of us? I was laughing when I told Jim that I think I know less about farming today then I did 30 years ago when I started out! We both had to laugh at that statement, but I was dead serious! As I look out across the water logged fields, and the year that lies before me. I think of my finances and wonder if the stuff I have yet to sell will cover all my expenses? I think of Ethan and Joseph my two hired men that are to me more like sons, than hired help. I know they won't always be here, and like each of my previous ones they too will someday leave here , and take a piece of my heart when they go! I know young people have no idea that they touch us older ones like they do! I think of the young man that possibly will come here in May and stay until November and realize if he comes he will be the most important crop that is cultivated here this year! I see all these things and realize that I have to give them all to our great triune God because I can't do a thing on my own without him!

> *John 15:5: I am the vine; you are the branches. If a man remains in me and I in him, he will bear much fruit; apart from me you can do nothing.*

So I guess it is another year I have to step out on faith, and know that God is going to take care of all of this, and me where he wants it all to go!

> ***2 Corinthians 5:7:*** *We live by faith, not by sight.*

What is faith any way one might ask?

> ***Hebrews 11:1:*** *Now faith is being sure of what we hope for and certain of what we do not see.*

Where does our faith come from any way?

> ***Hebrews 12:2:*** *Let us fix our eyes on Jesus, the author and perfecter of our faith,*

And why do we need faith anyway?

> ***Hebrews 11:6:*** *And without faith it is impossible to please God, because anyone who comes to him must believe he exists and that he rewards those who earnestly seek him.*

Good advice for us all, no matter what things may look like, or may seem right now!

Thro' many dangers, toils and snares I have already come.
'Tis grace hath bro't me safe thus far, And grace will lead me

home!

Your brother in Christ

George Denn

Added March 28th

On Thursday Joseph and I went over to help Wayne Schwartz cut down some trees out of an old fence line so Wayne can build a new fence. My chain saw became pinched in a tree, and when I got it out the chain wouldn't turn. Wayne and I were messing around with the saw trying to get it to work. Wayne had his pocket knife that he was cleaning the crud out of the saw bar with. After we got the saw running Wayne said he had lost his knife. We looked on the ground where we had used it, but we couldn't find it. Wayne was saying it should be easy to see, because it was so shiny. I told Wayne we should pray and ask God to help us find the knife, after we prayed; look as we may we just couldn't find it! Today I had to call Wayne to ask him some details about this story, because he told me yesterday that he had a miracle; he had found his knife! He was milking cows yesterday morning, he was checking to see what time it was, but when he reached inside his pocket to check his watch, the watch was not there, but the knife he had lost the day before was inside his pocket! Wayne said he had checked his pockets 2-3 times for the knife the day before, but it wasn't there and he never goes without his watch! I have had experiences something like that. Wayne's son Paul told me that his dad probably just misplaced his knife, but if he thought he had a miracle well good for him! I think I will side with Wayne because I saw him looking for the knife and I saw him check his pockets, and Paul well he was nowhere around when this happened!

Hey By George! April 26 2009

Matthew 14:13-21: *When Jesus heard what had happened, he withdrew by boat privately to a solitary place. Hearing of this, the crowds followed him on foot from the towns. When Jesus landed and saw a large crowd, he had compassion on them and healed there sick. As evening approached, the disciples came to him and said, "This is a remote place, and it's already getting late. Send the crowds away, so they can go to the villages and buy themselves some food." Jesus replied, "They do not need to go away. You go give them something to eat."*

"We have here only five loaves of bread and two fish," they answered. "Bring them here to me." he said. And he directed the people to sit down on the grass. Taking the five loaves and the two fish and looking up to heaven, he gave thanks and broke the loaves. Then he gave them to the disciples, and the disciples gave them to the people. They all ate and were satisfied, and the disciples picked up twelve basketfuls of broken pieces left over. The number of those who ate was about five thousand men, besides women and children.

I just had to share with you all the week I just had! Although it was not quite as dramatic as the scripture above none the less I felt I should write about it. Last Monday April 20 was my 47th birthday. About 9 AM my friend Tim Strommer called. He wanted to know if I would come down to where he works at about 4 PM that after noon to pray with him. I was telling Tim I would be there and that I could use some prayers myself, for a few things! So after an almost uneventful day I drove to Mankato to a place called Kids against Hunger. It is a place where food is packaged for places like Haiti and other impoverished places; it is the place my friend Tim manages. After a few minutes our friend Bob also showed up, he also had a few things he needed prayers for. After a short gab session we went into some prayer time. After our prayer time I was mentioning to the guys that no matter how tough our lives seemed right now we are still better off than the people in Haiti. A mud cookie that some try to eat to "feel full," and some pictures of the people down there that hang on the walls instantly tell me this! I was telling Tim that I just can't look at those people and the conditions that they live under without leaving some sort of donation. $5 was all I had but I had some food at home besides what will $5 buy you for groceries? I was thinking of those mud cookies, and I felt pretty blessed as I handed Tim my last $5. Bob was mentioning to me that he had some used clothes in his car that he would give me if I wanted them. "Sure," I said and I laughed, because anyone who has ever been to my house knows that I already have an abundance of clothes! I took them and thought maybe Joe my hired man can use them. As I was about to get into my pickup Bob mentioned that while in prayer the number 39 kept coming into his head. He said he looked in his wallet and he had thirty nine dollars in it. "Here George happy birthday," he said. As he handed me the money! "Many thanks Bob," I said as I gratefully accepted his gift! As I was backing out of the parking lot I heard a knock on my window I turned to see who it was and it was Tim. Tim said he was on a fast so he couldn't buy me lunch. I told Tim that I think Ethan and Joe were taking me out for Supper that evening. "Well here is $9 go and get something for your birthday," Tim said. I told Tim that I would and that evening I was able to buy $48 worth of groceries! It is amazing what God did with $5!

My friend Terry Tricky made the comment last winter while we were fishing at Lake of the Woods. "I want to be like you George and come home with more money than I went with!" I had to think about that statement for a while. I said. You know Terry it is true that a lot of times I have come home with more money than I went with; but there is part of the equation that you're not taking in consideration. Every time I have come home with more money than I have gone with; somewhere in the equation I had to give away all that I had! Something that is fairly easy with $5 but quite a bit harder with $500

After Monday my week went on; I had no sales at all, and by Friday things were getting fairly slim as far as my food situation again! All week I prayed for provision, but I could see none. In truth though God had already provided all I needed. I wasn't in want, but how many of us want to be uncomfortable? We like to have the things we need way before we need them. We don't like to have the things we need show up exactly when we need them; this causes us to be uncomfortable! Sometimes I feel God is not as concerned about our comfort as we are! Joe and I had been clearing some trees around a field that was shading out the crops. On Friday afternoon I pulled into my driveway with a load of wood. I thought I might as well get my mail. There was a birthday card in the mail with $20 in it. "I am rich!" Christine Tricky was cleaning my house again that day, and about 6:30 I was wanting a cup of coffee, but she was scrubbing the kitchen floor so there was no way I was going to get to the coffee pot! Just then my cell phone rang; it was my neighbor David Gibson wanting to know if I could give him a ride over to his dad's place so he could get his pickup truck. He had been planting corn over there and drove his tractor and planter back home so he had no way to get his pickup. If I would come now there would be a cup of coffee in the deal David said! I told David I would be right over, because I was pretty much done for the day with things around here! After a few cups of coffee David disappeared into his basement, he returned carrying a paper sack. "Do you need some hamburger" he asked? "I could use some" I said. What is interesting David had no idea I had fried the last pound of hamburger for dinner. There were 8 pounds in the sack! About $32 worth David Gave as a gift! The Holy Spirit sure must be working here was all I could think of!

When I returned home Christine was done cleaning, and was driving down my driveway. I stopped to talk to her, because I was wondering what the cleaning bill was going to be? "It's free George" Christine said. "God told me not to charge you anything before I even came out here!" I really thought this was interesting, because Christine was supposed to come here more than a month ago and clean; I had more than enough money set aside for this at that time. But I made an unexpected trip to Georgia, and Christine didn't want to come out here with no one here! So It got put off until now.

The next Morning my friend Terry stopped in. Usually we go to McDonalds for breakfast when he comes. Terry suggested because of the economic times that we are in; we should just go buy some groceries and bring them back here and make them. So we did and I spent $20.15 we came back here and started to fix breakfast. Joe was up when we got back and as we started to make breakfast Scott and his son Tony stopped by so we fixed some for them also, and we all had a great breakfast! I had at least half of what I had bought in town left over! While we were eating Scott was telling how he went a whole week without smoking! He sure looked better than he did the last time I saw him! Scott had been here a few weeks back. Terry, Joe and I were out in the woods cutting wood when Scott suddenly appeared! Scott was telling us that He was going to have to have back surgery and, because he was a smoker it would be more complicated. Terry and I prayed for Scott right there in the woods so it seems like the prayers have helped Scott!

> ***Jeremiah 29:11-13:*** *"For I know the plans I have for you," declares the Lord, "plans to prosper you and not to harm you, plans to give you a hope and a future. Then you will call upon me and come and pray to me, and I will listen to you. You will seek me and find me when you seek me with all of your heart." Well it sure looks like God knows what he is doing with us even in times we are not so sure! Gods' peace and abundant blessings to you all!*

Your brother in Christ

George Denn

Hey By George! July 4 2009

John 8:34: *Jesus replied, "I tell you the truth, everyone who sins is a slave to sin. Now a slave has no permanent place in the family, but a son belongs to it forever. So if the Son sets you free, you will be free indeed.*

WOW! Here it is the 4th of July already! Now I am as patriotic as the next person, but the 4th of July not only marks our independence from England, but to me a farmer from southern Minnesota it marks the end of the planting season and I start to look towards the harvest season. It is a time to watch my crops grow to see what they will become. Here are a few interesting events that happened around here this spring and just yesterday that I felt inspired to write about.

My harvest season actually starts around the first part of April. for the past 15 years God has blessed me with an abundance of fire wood that has either came in the form of dead trees, or trees fallen by the wind, or trees that just needed to be cut because they hang over fields and are in the way of farming practices. I use this wood to heat with and to sell during the winter. This year Joe, Ethan, and I were able to cut split and stack 120 cords of fire wood. One of my projects this year was to cut the trees around a 15 acre field that I rent about 3 miles from here, I have been procrastinating on this one for the past 20 years, but this year I had the help so I decided it was now or never! After we had been cutting for about a week my Landlords wife came out to see what we were doing, so I was telling her that I was trimming back the trees that had grown out on us over the years. She made no mention that she disliked what we were doing she just said that she had to get back to her yard work. On the tenth day of our cutting I estimated that we only had 2-3 days left on this place and we would be done. That evening while we were finishing our 22 load of wood my land lord pulls up in his pickup truck and says, "I have been hearing you cutting out here for the past 10 days and it makes me very angry. I want you to stop cutting right now finish what you're doing, pull the brush out of the woods that we had pushed in, I want half the money for this wood you're going to get, and if you cut one more tree I guess I will have to shot you!" He also wanted to talk about the rent this next Saturday, because he wanted more. Needless to say I was kind of stunned none of my other landlords had a problem when I had done this and some of them even encouraged me to do so! After our encounter he drove off, only to get stuck on a muddy patch of ground. I took my skid loader over there and pulled him out. He thanked me I told him no problem it was because of me that he was out there anyway. As I drove across the field on my way home I was thinking to myself that I really need to re-evaluate why I am farming this piece of land in the first place. When I first started farming this piece of land the previous owner came to see me, and asked me if I would grow hay on it. I needed the land at the time to produce the extra feed for my dairy cows. Now that I have no cows anymore and my life is about serving the Lord and not about making money I decided I would leave it up to the Lord whether I farm this piece of land anymore! One thing that puzzled me about the confrontation was why did he wait so long to come and talk to me, and why didn't his wife ask me to stop the day she was out to see us? I asked Joe what he thought about it. He didn't know he said. So I prayed about the situation. I figured if I could get the crop planted by the weekend I would farm it at least this year anyway. If not I guess I am out of here! I even took my sign down that I had on the place; just in case things didn't work out I would be already gone! We finished planting that field Friday evening May 1 just as it was getting dark. I was to go and talk with my land lord there the very next morning. How I hate confrontation! The next morning came I called him up and he said that we should meet around 10:30 am. I prayed and asked the Holy Spirit to go before me, and that he would be there during our talk. He mentioned he wanted more rent for the ground and half the fire wood this means I will be working over there this year for nothing! Yippee! When I got over there he was just pulling his pontoon boat out of the shed to go boating. We must have talked for at least an hour about many things. Then I said. "I suppose we better get down to business here. You mentioned that you needed more rent what amount were you thinking of?" He answered. "I was thinking maybe we should leave the rent the same as always just send me a check sometime." "How about the fire wood you mentioned you wanted half of that." I said. He said, "Just forget about that and I want you to finish cutting the trees back to where you were going to." He mentioned he was having a real bad day that day he confronted me and I was the one he took it out on! He also mentioned that it was his birthday that day he was 60 years old. I gave him a signed copy of my first book and wished him a happy birthday! How is that for a change of heart I am sure glad I didn't say much the day of our confrontation!

> ***Proverbs 27:19:*** *as water reflects a face, so a man's heart reflects the man.*

I am continually amazed by how much work gets done around here this year! So far by having 2 hired men around here I have been able to get things accomplished around here that I have been thinking about for years. It will be interesting to see what things look like around here when the year comes to a close and we see how things all end up, but I definitely can see the wisdom in Solomon's words!

> ***Ecclesiastes 4:9-12:*** *Two are better than one, because they have a good return for their work; If one falls down his friend can help him up. But pity the man who falls and has no one to help him up! Also, if two lie down together, they will keep warm. But how can one keep warm alone? Though one may be over powered, two can defend themselves. A cord of three strands is not quickly broken.*

Although having two hired men is great for getting the work done it has been quite a strain on the finances! But I know that God bought this equation together this spring and I continually see our triune God at work within our midst. More than once me and my guys were able to rescue a neighbor in need, who figured he could do more than he was able to do! I guess I have come to know my limits which are not much anymore, but they wouldn't be anything at all without Jesus in my life and the two hired guys he sent to help me this year! Just 2 days ago Joe and I was able to help my friend Wayne get his sorghum/Sudan grass planted, so he could go on a much needed trip this weekend! After I left Wayne's place I went to my dad's for a visit. I came home and worked around here until mid-night it just seemed like such a perfect evening! Yesterday morning my friend Terry came by. We were talking for a while and he asked," where is your Dodge pickup"? I said, "Isn't it in the front yard?" "No." he said, it isn't anywhere!" I just laughed. I like to say it just disappeared, Kind of like Enoch!

> ***Genesis 5:24:*** *Enoch walked with God; then he was no more, because God took him away.*

I don't believe my Dodge pickup was that good! As a matter of fact I am glad it is gone! Although there was many an adventure taken in this pickup with the Lord as our guide, it was a financially evil piece of metal! Not only did it need an oil change severely it also needed some $3000 worth of repair mainly on the front end. Not to mention the $2800 that I still owed on it yet! Being three months late on the payment might shed some light on where it went to! But I am not that curious! I knew I was going to have to have something better than my old red ford to get by with around here. I was telling Terry I knew where there was an older ford pickup for sale; just up the road a bit, but until some funds came in I just couldn't buy anything. Terry was saying that we should go and look at it. I said to Terry that I didn't want to because I didn't have any money. Terry said that we should go and look at it and that maybe he could help me out with buying it. I said, "Well ok I guess it wouldn't hurt to look." The pickup was a 1983 gold f150 ford it had 57,000 actual miles on it my neighbor was the second owner he had bought it from a retired farmer. My neighbor just had put in new brakes all around and new front rotors he wanted $1200 but Terry talked him down to $1000 cash money. I was telling Terry that I had noticed a black Ford pickup for sale last night at the end of the street where my dad lives up in Waseca, and that being Terry had to go there to get the money we should look at that pickup first before we decide. The Black Ford pickup turned out to be a 1989 f150 4x4 it had 137,000 miles on it Terry knew the guy who owned it he was a mechanic and he had just went threw everything in the front end! He wanted $1600, but Terry talked him down to $1200 cash! Terry was saying that maybe I should buy both pickups with the pumpkin season not too far away. Terry said he had the money and he wouldn't charge me any interest and I could pay him back when I could. Terry had the money hid in a mason jar saved for his daughter Elisabeth's college funding! After we made the transaction on the black Ford a man called and told me he had just left a check for some hay he had bought a week ago, so I told Terry I would be able to pay him back half of the money for the pickups that very day! Terry was kind of glad, because his wife was kind of upset with him for helping me out after I said the Dodge disappeared! We got home and went and bought the other pickup as well. Both pickups are really sound I wouldn't be afraid to drive them anywhere! So now I am all set for the time being, so all's well that ends well!Here are some interesting points to the story. The front end and brakes were in need of repair on the Dodge. On both pickups I just bought both front ends and brakes were just put in. The Dodge needed an oil change. Both pickups had just had there oil changed! The miles on both pickups put together add up to almost exactly what the Dodge had on it! 195,000 miles! The Dodge had a half tank of gas in it the 1989 Ford had between a half and three quarters of a tank of gas in it! So all I can say is thank God and for his Holy Spirit working through us all! And thank God for Terry's Mason jar with all the $100 bills not only will his daughter Elisabeth be able to go to college because of it, but a lot of us learned a few things as well! Happy Independence Day!Galatians 5:1 it is for freedom that Christ has set us free. Stand firm, then, and do not let yourselves be burdened again by a yoke of slavery.Gods' peace and abundant blessings to you all!

Your brother in Christ,

George Denn

Hey By George! August 4, 2009

Romans 8:16-21: *The Spirit himself testifies with our spirit that we are God's children. Now if we are children, then we are heirs-heirs of God and co-heirs with Christ, if indeed we share in his sufferings in order that we may also share in his glory. I consider that our present sufferings are not worth comparing with the glory that will be revealed in us. The creation waits in eager expectation for the sons of God to be revealed. For the creation was subjected to frustration, not by its own choice, but by the will of the one who subjected it, in hope that the creation itself will be liberated from its bondage to decay and brought into the glorious freedom of the children of God.*

I just returned home once again from 2 weeks of summer camp. Before I get too involved with the wheat and hay that needs harvesting as soon as the weather is right. I just wanted to share my experiences of the past 2 weeks at camp!

Every year for the past 6 years, I have been involved with summer camp. At first it was just one every year. Now I have 2 camps that are back to back in July. I have to juggle my work load to do this, but with a whole lot of help from the Almighty it has been possible. As usual there seems to be a force that tries to prevent me from taking the time off to do this! My neighbor David never seems to fail to mention that I should just stay home and take care of the farm, and then I wouldn't have so much commotion in my life! This may be true, but then I wouldn't have all these experiences to write about then would I? The one thing I had to finish before I left was the second crop of hay. We started cutting it on July 6, so I figured I would have plenty of time before I had to be at Northern Light camp on the 18th. Here are some of the things that happened during this hay making episode! Not only did we manage to wreak my haybine but also destroyed a shaft on my neighbors haybine. It should take about $6000 to fix mine and $900 to repair my neighbors! We had three hayrack stringers break and one running gear that just "collapsed" as Joe put it! He just turned to short with it; I think I forgot to tell him about that! My skid loader had a busted engine ring and a wheel bearing that was out, but I took it easy with the thing and we did get done. Howard Guse hauled it over to Wayne Schwartz to repair for me on payment for some hay. The rear engine seal went out on the 89 ford, and we hit a deer going home one evening with the 81 ford! On the bright side, God provided 3 Amish men to buy the hay so I could even go to camp! Their Names are Allen, Toby, and Dan. I always enjoy talking with the Amish. Howard Guse and I were down by St. Charles, Minnesota about 110 miles east of me delivering a load of hay to them. The day we were at Allen's place, I found myself up in the hay loft, moving hay around with Allen's sons, Norman and Freeman. I found out that Norman was 16 and out of school for 2 years. Freeman is 12 and has 2 more years of school to go. I think I would have liked to have been Amish when I was a boy! But I think now the horse thing would be a real drag. Maybe if someone else would shovel the poop and get them all harnessed and bring them up to the door I might think about it. Allen was telling me that in his opinion it's a shame. When he goes to town and sees these 15 year old boys, just playing ball, when they could have some cows to milk! I just laughed and said "yea I don't think that it would hurt them any"! Allen is 39, and has 11 children and farms 90 acres.

I arrived at Eagle Bluff where Northern lights camp is held, a full hour late. While in thought I missed a turn and went a different route. I came upon road construction Just about everywhere! So I was glad to arrive at my destination. I was Counselor for the 15- 18 year old boys. This included Austin, Erik, Andrew, Joshua, and Coulter. Mike Haack was my assistant, and my first answer too prayer! I had some concerns about a few personalities' that would be in my dorm. Mike is a big man and no one messes with him! The theme of both camps was centered around the movie The Chronicles of Narnia. The Witch, the Lion, and the Wardrobe. And a scripture from

> ***2 Corinthians 4:18:*** *so we fix or eyes not on what is seen, but what is unseen. For what is seen is temporary, but what is unseen is eternal.*

The first night was so very late that I didn't do dorm chats. I was sort of glad because I didn't quite know how to take the group with conversation that late the first night any way. At first it just seemed that these five guys just wanted to try me to see what they could get by with. I just told them that no matter what they we just playing into my hand! On Monday my guys had Archery and Group challenges for activities and after chapel in the evening we finally was able to have a dorm chat that evening. I invited Steve Deuel to join us also in dorm chats. I have long understood that there is strength in numbers! Before we went into the room Mike, Steve, and I stood outside the door. After making some sort of comment, about the jabber that was coming from inside our dorm. We decided that it would be best to pray before we went in! Upon entering the dorm the Holy Spirit told me to go to them on their level. So we adults sat on the floor in their room. I think I asked the question what each of us wanted to take home with us from this camp? God just totally took over our dorm chat time and ended our session with prayer. I was totally amazed! On Tuesday my guys had Minimum impact camping and earth exploration. I made a dream catcher in Betty Johannsen's art class in the evening. On Wednesday we had rock climbing and Gps Path finders. On Thursday we visited an Oneota Indian village from 1000 years ago and went river canoeing in the afternoon Erik was my canoe partner and things just seemed to go better when it was my turn to steer! Friday we did a tree tops course or I should say they did. I seem to be less interested in heights the older I get. It rained that afternoon so we didn't get to do the pioneer village activity of coarse the one I was looking forward too! We had Chapel every evening around 7 o'clock. The messages were the same for both camps they are as follows.

Chapel 1 message a search for shelter.

Main point~ Remember that you have a place and someone to turn to for love and protection.

> ***Psalm 18:2*** *The Lord is my rock, my fortress and my deliverer; my God is my rock, in whom I take refuge. He is my shield and the horn of my salvation, my stronghold.*

> ***Matthew 7:25-26:*** *The rain came down, the streams rose, and the winds blew and beat against that house; yet it did not*

fall, because it had its foundation on the rock. But everyone who hears these words of mine and does not put them into practice is like a foolish man who built his house on the sand.

Chapel message 2 what should you believe?

Main point~ Remember there is more to believe in than the "UNREAL" world that we see.

1 Corinthians 2:9: *However it is written "No eye has seen, no ear has heard, no mind has conceived what God has prepared for those who love him"*

2 Corinthians 4:17-18: *for our light and momentary troubles are achieving for us an eternal glory that far outweighs them all. So we fix our eyes not on what is seen, but on what is unseen. For what is seen is temporary, but what is unseen is eternal.*

Chapel 3 God is preparing you to be an heir.

Main point~ all though we may feel like we are UN worthy to walk with God. He is using our current situations to prepare us to do great things!

1 Samuel 16:7: *But the Lord said to Samuel, "Do not consider his appearance or his height, for I have rejected him. The Lord does not look at the things man looks at. Man looks at the outward appearance, but the Lord looks at the heart."*

Philippians 4:13: *I can do everything through him who gives me strength.*

Chapel 4 who will you believe?

Main points~ there are multiple voices in your life who will you listen to?

Jeremiah 29:11: *for I know the plans I have for you," declares the Lord," plans to prosper you and not to harm you, plans to give you hope and a future.*

Chapel 5 the ultimate sacrifice

Main point~ there is more to believe in than the "UNREAL" world that we see

> ***John 3:16*** *:"For God so loved the world that he gave his one and only Son, that whoever believes in him shall not perish but have eternal life.*

Chapel 6 which reality will you choose?

Main point~ Remember that you do have a choice which reality will you choose?

> ***2 Corinthians 4:18:*** *So we fix our eyes not on what is seen, but what is unseen, for what is seen is temporary, but what is unseen is eternal.*

Throughout the week I felt that our dorm had bonded. we didn't give out awards, but I felt my guys deserved something so I bought them all a snickers candy bar from the camp store, and pronounced them all "TOP GEZER'S" a quote from the movie we had watched earlier in the week! Colter made the statement that he was going to retire me soon and then he would be the counselor! I said good because I was ready to be retired! Besides one incident on Friday evening, in which was made right on Saturday before camp ended in my dorm. I will say was a very successful camp and ended on a positive note! I always have mixed feelings at the end of camp sad to have to leave my guys that I worked with that week, and to see them grow spiritually, but also happy to be relived of the responsibility because it is a big one and I don't take it lightly! After saying Good bye to the guys in my dorm and to the staff that I worked with it was time for me to leave for my next camp. Erik helped me load my things I gave him a big hug good bye. I think I cried for a few miles down the road! I went south through Preston and Harmony Minnesota. Again I saw the Amish buggies in town amongst the cars, and I remember the Amish who made it possible for me to even be on this trip! I drove south through Iowa and noticed a big stretch of hailed out fields. I was wondering if I had gotten any back home but resisted the urge to call home. Besides it would bother me all week if I had gotten any! So I didn't call. I figured Ethan and Joe had everything under control, and they would call me if they needed to anyway! After the 6 hour drive the first person at camp that I said hi to was Erik Wright. I thought that was interesting because the last person I said goodbye to up in Minnesota was Erik from my dorm! My duties at Heartland camp was security and maintenance. The first thing I did was to spray all the wasp nest around all the buildings. Of course there was the usual card game up in the dining hall every evening! There was a camper, who remembered me, but for the life of me I couldn't remember his, and he seemed disappointed. Rex was his name and I told Rex that by the end of camp I was going to make a point of remembering his name. Which I did! It was also very rewarding to see several of the guys that were on staff as co- counselors and that had other positions around the camp, were once in my dorms when they were campers! One evening while we were taking communion, and after I had returned to my seat. I was watching all the staff and campers that I had come to know over the years, and had several of the young staff and campers pat me on the back as they walked past behind me, after they had taken there portion of the communion. It truly was a unique experience! I could really see Jesus working in and through them! One evening we had camo night the youth were given instructions about the course. Some staff like me was put in strategic places to give the youth miss information and some were to give them right information. If a camper passed by me I would say nothing and if one asked for advice, I would tell them "Go to the camp store." If they heeded my advice they would find only darkness and water balloons thrown at them if they lingered too long there! It was a very interesting night and I thought how similar life was to this game. God was one of the characters, played by David Howe. He was up on a roof shining his light to light people's path to show them the way, but most didn't want to go into the light!

John 3 19:21: *"This is the verdict; Light has come into the world, but men loved darkness instead of light because their deeds were evil. Everyone who does evil hates the light, and* *will not come into the light for fear that his deeds will be exposed. But whoever lives by the truth comes into the light, so that it may be seen plainly that what he has done has been done through God."*

The Holy Spirit was walking around, pointing the way that they should go, but the Devil was also walking around convincing youth to follow his deceiving advice, and most did at first, only to be pelted by water balloons in the end! A light was shining at the big white cross that is there at the camp. The campers had been instructed that all truth could be received freely at the cross! I took some time, but eventually most of the campers found out that they had to visit the cross, if they were to finish this game! And soon I was out of business. Two older boys that evening ran head long into one another, and had to be taken to the hospital. Before the incident they could barely stand to be around one another, but afterward they were the best of friends! One afternoon I stopped in on Richard Bengtston's pottery class and with some help from Richard I was able to make a pot, but I found out as most did it is not as easy as Richard makes it look! On Wednesday evening they had a dance in the cafeteria. Dave Salanders and I were out patrolling the grounds in the Gator. I called the one we were in the "granny Gator" as it had no power, and had smooth tires! We were out at the look out on top of the steep hill. I asked Dave if he thought we could make it down the hill on the Gator. He jumped on and we started down. The Gator started to free wheel and the brake would only hold on one wheel. We started to fish tail and then I lost control, and we twirled helplessly down the hill! At one point I thought the thing was going to flip, but the grass was already getting wet from the dew and the tires were smooth so we made it to the bottom unscathed! After a good hearty laugh we drove back to the dance. On the way back I thanked God for protecting us, and we had a good story to tell when we got back to the others! At the end of camp 12 campers out of 80 decided to be baptized. So I would say camp ended on a successful note! As It got time to leave camp I said my goodbyes to those I had worked with. I knew some I may never see again! I hope to go back to both camps next summer God willing! Two weeks is a long time to be gone from here during the growing season, but the rewards I receive from being at these camps can't be measured with anything monetary! As I drove east on Ill. hwy 17 to go and pick up my father in Momence. The song we sung at the end of chapel several times came into my head.

Jesus Messiah
Name above all names
Blessed Redeemer
Immanuel
The rescue for sinners
The ransom from heaven
Jesus Messiah
Lord of all!

And when I returned home, I found out that it had hailed while I was gone. The boys didn't want to tell me, and let it bother me while I was gone. However I feel God protected the wheat crop and the pumpkins only 6 acres of corn looked bad. So I guess I can handle that!

2 Corinthians 3:17-18: *Now the Lord is the Spirit, and where the Spirit of the Lord is, there is freedom. And we who with unveiled faces all reflect the Lords glory, which comes from the Lord, who is Spirit.*

Gods' peace and abundant blessings to you all!

Your brother in Christ,

George Denn

Luke 15:8-9: *"Or suppose a woman has ten silver coins and losses one. Does she not light a lamp, sweep the house and search carefully until she finds it? And when she finds it, she calls her friends and neighbors together and says, "Rejoice with me; I have found my lost coin.' In the same way, I tell you, there is rejoicing in the presence of the angels of God over one sinner who repents."*

Today I found the old tractor crank! I have been looking for it for almost seven years, but the thing just seemed to have disappeared! The morning had started out ordinary enough, and after 2 months of selling pumpkins, and October being the rainiest and coldest one on record. I had been feeling spiritually flat! Just like a flat tire that can't be moved until it is pumped up! Just the other day I had been telling my friend Terry that I didn't think that I had any faith left anymore. And probably the times that I thought I had any faith. I was probably just kidding myself! I thought too, that it had been a long time since God had inspired me to write about anything. Perhaps he was done using me in that way? Or so I thought.

Today we spent the morning taking down the pumpkin stand that is south of Mankato on hwy 22 on Al Schenk's 40 acres. Around noon I sent Joe over to rake 90 acres of hay that had been cut since Oct. 3. The plan was that while Joe raked the hay, David and John and Troy and I would go over to the pumpkin stand on hwy 60 near Elysian and take one down as well this afternoon. Joe called me around 2 o'clock and told me that the rake had busted in two right behind the front wheel! I told Joe to hook up the other rake that I have, and to continue on, but in a different fashion. I told the others that I would meet them at the stand after I went to see how Joe was doing. I stopped to look at the busted rake; he was right it had busted in two! I stopped to tell Joe that it wasn't as bad as it seemed. He asked me,' "how worse can it get than being broke in two? "Maybe I am just getting numb to it all! It all looked like a simple fix to me; these young guys seem to get excited about everything! I also told Joe that I thought it was getting to windy to rake hay today and that he should come over to the stand and help us take it down. We loaded up the pumpkins that I was going to save for seed, and hooked up the hay racks that were there. I was going to stop and get another of my racks from a guy that had one of mine who lived not too far from there. I was looking for another hitch pin on the floor on the passenger side of my black 89 half ton Ford. It was the one I mostly drove all fall. I was digging around for a pin in a pile of accumulated stuff. When I pulled it out of the pile I could hardly believe my eyes! On the floor of a pickup that I only owned since July; was the crank for an old WD 45 Allis Chalmers tractor that I once owned! I had not seen this crank for at least 7 years. I often wondered where I had put it, because I wanted to keep it for a souvenir. This crank was special. In the fall of 1994 I was drying corn; it could be quite possible it was on this very day! The old 45 Allis I was using on a grain elevator wouldn't start, so I grabbed the crank to start it! Being younger, and more agile, less patent and farther away from God than I am now. I imagine I had a few fine words mixed with some tobacco juice, as I put the crank in its place and with both hands gave its motor a spin that should have started anything! The thing backfired and the crank spun around backwards and hit my left arm and knocked me to the ground! My arm felt like it had been hit by something that was out of this world! I got up swearing and grabbed the crank, but I noticed it was bent out of shape when it hit my arm! I have never been able to crank a tractor since it's like a fear that I have now! I remember taking the starter out to be fixed immediately after that happened. I was telling the story what had happened at the implement were I went to get a new starter. The older guys who worked there at the time seemed amazed, and mentioned that they never heard such a thing without the guys arm being broke in the proses! I will always believe it was the Lord that protected me from my arm from being broken that time! As David Kruse and I went to retrieve my other hay rack. I was telling him the story about the crank. After I told the story he was asking me what it meant to crank a tractor. A question not only did I find humorous, but also something tells me I'm no spring chicken anymore! I found it interesting also when we got to the guys place, he had a tractor exactly like the one I was talking about hooked to a grain elevator unloading corn and my old elevator that I once owned at the time, but he had since purchased was sitting off to the side! Hmm very interesting! I was telling the rest of the guys back home the story and they seemed pretty amazed by it I was even able to find another crank in the shed so they could see what it was originally supposed to look like! Still nobody seemed to know where the crank came from? Troy had not been present when the story was being told so I asked him if he knew anything about it. He said a couple of weeks ago he had been out to pick up some corn bundles out in the field. When the truck stopped and Troy got out he said there was this piece of iron laying right there on top of the ground. He picked it up and was going to throw it into the old pasture when he had a thought that it looked like an old crank for a tractor or something, and perhaps I could use it. So he put it in the pickup and forgot about it until now! That was the fall before I started my spiritual journey. As the snow falls lightly outside my window as I finish up this story, and once again the fields are covered with snow, another year is coming to a close. I am reminded of how I was back then at age 32, and were I am now at age 47. I think that Solomon was very wise when he wrote.

> ***Ecclesiastes 11:9-10:*** *Be happy, young man, while you are young, and let your heart give you joy in the days of your youth. Follow the ways of your heart and whatever your eyes see, but know that for all these things God will bring you to judgment. So then, banish anxiety from your heart and cast off the troubles of your body, for youth and vigor are meaningless.*

I also feel the pains of growing older and finance's less then I would like to see after so many years of work! I am led to the scripture in Psalm 9:10:

> *Those who know your name will trust in you, for you, Lord, have never forsaken those who seek you.*

Well I have to admit after all this time of following Jesus he knows what he is doing in my life even if I can't see him, and even at times when I feel I have no faith at all. Sometimes I have to re-read the story's I have written and recall some that have not been written about just yet!

Philippians 1:6: *Being confident of this, that he who began a good work in you will carry it on to completion until the day of Christ Jesus.*

I guess he will do this even if he has to use a tractor crank to get you started!

Gods' peace and abundant blessings to you all!

Your brother in Christ,

George Denn

Hey By George! April 12, 2010

> ***Matthew 13:3-9:*** *Then he told them many things in parables, saying "A farmer went out to sow his seed. As he scattered his seed, some fell along the path, and the birds came and ate it up. Some fell on rocky places, where it did not have much soil. It sprang up quickly, because the soil was shallow. But when the sun came up the, the plants were scorched, and they withered because they had no root. Other seed fell among thorns, which grew up and choked the plants. Still other seed fell on good soil, where it produced a crop-a hundred, sixty or thirty times what was sown. He, who has ears, let him hear."*

After living on this place for almost 48 years now, I can start to see some patterns forming! I came to see this year that it seems once in every ten years, I am able to be out in the fields in March. I had planted oats and wheat in March in 2000 and also was able to plant wheat in March in 1990. Usually when this happens in southern Minnesota we have had a dry year previously and not much snow during the winter. This time that was not so! We had about the worst Oct. I think it rained for 22 days, and we had a lot of snow and fairly cold winter. So when Donna, my dad's cousin came to visit on the first part of March. I had no problem agreeing to take her back to her home in Illinois around the 24th. I hadn't expected the snow to be all melted and the fields to be almost dry to plant wheat! So I asked Donna if she wouldn't mind going home a day sooner, like on the 23rd. This trip would take a few days, and when I got back I figured I could probably start planting wheat. I was also contemplating a 9 day trip to Grove, Oklahoma. To visit Brad Claggett and back across to Rolla, Missouri. Where I would stop off to visit my friend Michael Yegerlehner and hopefully see Jon and Carla Reinagel as they were back from Africa for a short time. After this I would go to some camp training up near Peoria Illinois for the weekend. If God would make this all possible! My hired man Joe Kruse had just bought a 1993 Honda Accord. I asked Joe if I might use his car for my trip. He said yes," but sometimes it made him wonder. "I told him it wouldn't be any different than when he was using my stuff last fall and he had the field cultivator climbing over the tractor wheel. Or last summer when he called and told me that a hay rack had just collapsed! Well it looked like it all depended on the wheat. If I got it planted or not! If I didn't get it planted, I really didn't want to be gone thinking about it all for 9 days! Is this possible Lord can you help me with this?

> ***Mark 9: 23-24:*** *"'If you can?' said Jesus. "Everything is possible for him who believes." Immediately the boy's father exclaimed. "I do believe; help me overcome my unbelief!"*

Well here is how it all turned out! I am warning you just don't blink! I returned home on Thursday the 25th of March from taking Donna back to Illinois at about 2:30 pm. I had seen some field work being done on my way home. So the first thing I did was walk across the field here at home where I was going to plant the wheat! It looked dry enough to work so I needed to go get some fuel I already had the field cultivator hooked up to the tractor. I put some fuel in and out to the field I went! I was about 1/4 done with the field when I got stuck in some mud. The only way I could get out myself was to unhook the cultivator from the tractor and pull it out with a log chain! After this process was complete I hooked it back up and continued my work. Again I got stuck, again I had to UN hook the cultivator, but this time I couldn't just pull it out. I had to hook a chain over the top so the cultivator would just flip up and over so I could drag it out of the mud, and then repeat that process so it would be right side up! Keep in mind this implement is 26 feet long and 10 feet wide. Needless to say it was quite a site! It's the kind of thing you're glad that no one can see, because I think people wonder about me the way it is! After I got hooked back up I continued on, but once again I got stuck, and you know what I did? Wrong! I left the cultivator sitting in the mud hole. I had enough for one day! As I was lying in bed that evening I remember mentioning to God that if that's the way it was all going to be this spring I think I will wait a few more days! Of course there is always rain in the forecast when I am trying to get stuff done and everyone seems to have to go out of their way to tell me "It is supposed to rain!" I got up on the 26th with renewed hope. I was going to seed some red clover on 7 acres that had been planted to rye last fall. Then I had to go to town to get a couple of new log chains, the ones I had were all broken and stretched from a tree stump pulling episode last fall! I came home and did the double flip thing with the field cultivator. That was the last time I got stuck and ended up finishing around 3:00 PM. I started planting and by the time it got dark I was almost done so I kept going and I finished around 8:30 PM. I could tell that God was helping me out because it sure seemed like I had got a lot accomplished this day more than I should have! I did have to leave a few wet spots, and of course it never did rain. All that aggravation for nothing! My next field was 22 acres and 12 miles away. I drove the tractor over there on Sunday the 28th of March, and got stuck the ground was really soft. I decided not to plant it before I left on Friday, unless God miraculously dried it out by then. In some ways I thought he was doing just that the wind blew hard all week, but by Wednesday evening I had to pull back home. I could till the spots I had left last week, and planted that on Thursday around 4 PM. We cleaned the grain drill out so we would be ready to plant the 22 acres when I got back home. I promised myself not to think about the 22 acres that had not gotten planted while I was gone, and I didn't! On Friday April 2 at 6:00 AM I put the farm on Kruse control, and put the car on cruise control, and headed for northeastern Oklahoma! My first stop was to Hy- Vee grocery store in Mankato to get some grapes and veggies to eat while I drive. I am trying to lose some weight. A doctor told me I was about 60 lbs. over weight. She said if I was going to pig out I should do it on fruits and veggies. So instead of potato chips it is now grapes and cauliflower! I noticed that there was not much field work being done as I drove. Every once in a while I could see the remainder of a snow bank in the ditch as far south as Des Moines Iowa. Somewhere south of there a huge thunder storm came up. To the left of me the sun was shining, but on my right the sky was black and stormy! When it hit I was just pulling out of a wayside rest and had to drive pretty slow and in the right lane on I-35. Some people had pulled over, but in time the rain slowed and you could see again! Somewhere close to the Iowa/Missouri border I needed to stop for gas, but the place I pulled into had no electricity so I had to go farther down the road. Not until I reached Bethany, Missouri was there any electricity. If I couldn't have gotten gas there I would have stayed until I could have, because my tank was empty! Some were I have heard that the name Bethany means the house of bread; it is a place I usually stop, when I am in the area! After I left Bethany my next Hurdle was to drive through Kansas City I just hate driving through there. I was asking God to please not let it rain will I was going through there! God did answer my prayer but as soon as I was going south on hi way 71 it started to rain this time it was sunny to my right and black off to my left by the time I got to Joplin it had quite raining. As I drove through the last of Missouri and into Oklahoma I was marveling how green everything was down there! I was to learn that spring had come later down there and earlier for us so they weren't too far ahead of us maybe a couple of weeks. I also noticed that it was extremely wet south of Des Moines. Hardly any field work had been done. Brad told me that everyone was going over to his Grandparents this evening so I should just drive over there. Kermit and Martha Claggett live just a few miles west of Vinita Oklahoma. On the corner of hwy 66 and 60 exactly 625 miles from my place. It sure was a treat to see and talk with these folks again! I think it's been 4-5 years since I've wandered down that way. Kermit was out in the yard when I pulled in. I told him that he probably didn't remember me that I was a friend of Brads from Minnesota. "Yes." He said I remember you. Kermit said that we might just as well go into the house and talk. Now Martha keeps a pretty clean house, so when I saw her look at my shoes after she gave me a big hug; I sort of laughed to myself. I had just washed them yesterday morning because they had dried mud on all over! So I was safe to enter! The next morning I followed Brad back to where he lives on a small farm near Grove Oklahoma. About 30 miles south east of Vinita. Brads Dad David and David's business partner own the farm. Brad is planning on raising some pumpkins down there. I gave him a coffee can full of seeds that I saved from last year's crop. For a joke I took some Minnesota dirt in a 5 gallon pail and gave it to Brad and told him that now he had some good soil down there! Brads place seamed peaceful, and I was very comfortable there. Brad has a bass boat he bought last year. One day we were going to go fishing, but when we pulled out of the drive and onto the road one of the tires started making a slapping sound because the rubber was leaving go from the tire. So we spent most of the morning in Grove getting a tire fixed. Brad showed me where his dad David and sister Tishra live. A lady was working in the yard next door Brad went over to chat with her. Her name is Mary I found out she is writing a book about living through the dust bowl days in Oklahoma! I asked Mary if she thought that young people of today could live through something like that. Mary thought before she answered my question and replied. She thought that they would find it within them to make it through just like the people of her time did. She did feel that it had robbed her of her teenage years though. She also said that Movies that were made like Grapes of Wrath did not portray the people in a right way. "We were just honest hard working people caught in a bad situation it just wouldn't rain! The rains did come back, but before they did I think I heard somewhere as much as 36 inches of topsoil had blown away! And some of it blew as far east as the Atlantic ocean where they found traces of Oklahoma dirt on the decks of ships! I saw a tree tipped over in a woods on Brads farm in its roots was just a few inches of soil some gravel and solid sand stone all in that order. Most everything is grass in that area to hold what little soil is left, but occasionally you see some plowed ground. Later that afternoon we did get the boat launched, but I got a sliver of glass in my foot in the process, and we didn't get too far before Brad found out that he forgot to put the plug in the boat. We did make it back to the boat landing before we totally sank! We had to load the boat up to put the plug in so we decided that we would go back to the house we had enough for one day! The next couple of days the wind was blowing too hard so we couldn't take the boat out. So we moved some old machinery and a couple of old pickups so they were out of view, and straightened out a shed so Brad could use it more efficiently. The last day I was there we did get to go fishing, but they weren't biting. So we went home! That afternoon we went to get a ladder at David's to trim some trees. Josh Zigler came out to help us. I met Josh last winter on a fishing retreat. Josh and Brad were riding a snowmobile for the first time and Josh found out you can't hit a snow drift sideways at 40 miles an hour! I was out trimming trees and I ended up cutting off a rose bush that Kelly wanted to keep! Everything looked so good she thought that she didn't mind the rose bush being eliminated! On Thursday morning just before I left Oklahoma David took Brad and I to visit a place outside of town that has been experimenting with Humates. A type of organic fertilizer, for 40 years now. The unprocessed humates looks a lot like coal. It's not as old as coal therefore it has a lot more nutrients in it. They took us through their small possessing plant just outside of Grove. I won't elaborate here but the people who own this facility are also using it as a ministry. Helping people in places like Africa to grow their own food! It's kind of like the concept of give a man a fish and you feed him for a day teach a man how to fish and you feed him for a lifetime! I had it in mind to leave Brad's place about noon on the 8th. When I am traveling I like to keep to my schedule as close as I possibly can. I always think about the story in Judges 19 how this Levite lingers longer than he intended too at his concubines' fathers' house, and what happens as a result of staying longer than he should have! I will let you read about it sometime on your own so you know what I mean. After I left Oklahoma my next stop was to be in Rolla Missouri. After slightly less than a 4 hour drive I arrived in Rolla just a little after 4 PM. Michael had said that it might be hard to find a spot to park near his house so I should just look around. As I was looking for Michael's house I drove right past it about the same time I found it! I drove down the street until I found a spot to turn around and I was able to park right out in front of the house! I could see that as a God thing! It was good to see Michael again! He and I worked at S.E.P. camp a few years back Michael was our dorm councilor and I was the co-councilor. I hadn't seen Michael for about 11/2 years now, he and Jon and Carla Reinagel were up to my place about Christmas time in 2008. I was hoping to see Jon and Carla also on this trip but it wasn't meant to be they had to leave for some family gathering a day before I arrived. Michael showed me around the house he rented, and shared with 5 other guys they are all in college Michael is finished and looking for a job as a teacher. Each person has their own room, and they share a common living room, and kitchen, and bathroom. I told Michael that I finally found a house that was in worse shape than mine! I know some people I know might have a hard time believing that, but now I have proof! Michael took me for a walk around the campus, and showed me various buildings, and what they were used for. There were a couple of quotes from Mark Twain and Albert Einstein written on the sidewalks. They reminded me of a quote that hung in my high school "he who fails to plan, plans to fail." I always wondered who on earth wrote that because it never seemed to work that way for me. WC Fields had a quote also "If at first you don't succeed try try again! Then quit there's no use being a darn fool about it!" I guess when your famous you can say whatever you want and people will run with it! I was telling Michael as we walked this was about as closed to college as I had ever come. Other than the 10 days that I spent at the Nashville Auction School. I pretty much am a student of the school of hard knocks where every lesson cost you dearly and is never forgotten! When I was in high school I was of the mentality that if I could add and subtract, and read and wright I would be able to make it in this world. I was doing pretty well until computers came along and messed everything up on me! When we got back to Michael's house he was showing me his old Grand Torino. I was telling Michael that I used to ride to school in one just like it. My friend Butch used to pick me up every morning I think I paid for some of the gas. That was 32 years ago and gas was 50 cents a gallon I think Butches car was in better shape too! This old car was parked under a roof that was sagging because one of the posts that held it up had fallen down. I propped the post back under the roof. I told Michael that should hold for a while and Michael made a comment about his "slum lord!" We both just laughed and decided it was time to eat so we went to a steak house/buffet. It sure seems impossible to lose weight while visiting friends! After we ate we went back to Michael's so we could make a map for my final part of my journey the next day. Michael pulled up a site on his computer that showed a picture of his house. We found my farm also, 435 miles away as the crow flies! It must have been a pre-2005 picture because the two cattle barns that were here were in the picture. After that we went to Denny's for coffee. After that we found a movie to take back to Michael's to watch. The next morning Michael made breakfast

Michael's coffee pot was a slow outfit! It only makes one cup at a time, because the decanter broke. It took me some time to fill my thermos! Michael made Italian spaghetti for dinner then it was time for me to leave. We prayed for each other before I left. I sure appreciate the friends God has given me. I am reminded of an old song we used to sing back in grade school. "Make new friends but keep the old one is silver and the other is gold." So long Michael till we meet again God willing! I had to drive thru St. Louis Missouri which to me is even worse than Kansas City! I had been through there several times, and I think the last time I was driving through there Jeff was with me and a super bowl game had just finished what a trip! Both times! I was watching the signs for I-55 north. I think Satan drives an 18 wheeler, because exactly when I needed to see the sign this semi pulls in front of me, and I went driving by my turn! Do you know what I said when this happened? Well I'm not going to say because it isn't very nice! Once I got onto 55 north I got myself into another pickle. I was supposed to get into the right lane as the road was changing directions. I didn't have much time so I force my little Honda Accord in between 2 semi's. That move made a few semi horns blow! But I refused to look at them and give the angry semi drivers any satisfaction! Finally I did get through St. Louis unscathed! A friend of mine Randy tried to call me while I was driving through St. Louis, but I thought it best not to be on the phone until I was out of the city. I made my way to the Salvation Army camp, Eagle Crest. Where we hold our spiritual enrichment program.(S.E.P.) Which is 10 miles south of Lacon Illinois on hi way 26. I noticed some field work going on as I neared Peoria, and I tried to remember that I was on vacation, and not think about the wheat that I needed to plant yet! It sure was a great treat to see all those friends that I have come to know from camp! Some of us sure get into some wild card games. I think we have come to call our game extreme ucker. Also it's just good to catch up on what's going on in everyone's life. On Saturday we spent most of the day in training sessions learning how important it is to work as a team! One of the really great things too see for those of us that have had staff positions, for any amount of time is too see young people come on as staff who were formerly campers. On Saturday evening we stopped our card game immediately to listen to one of our brothers share some problems he was having. The seven guys that were at the table have committed to pray for this person as well as one another.

> ***Galatians 6:2-3:*** *Carry each other's burdens, and in this way you will fulfill the law of Christ. If anyone thinks he is something when he is nothing, he deceives himself.*

It is interesting what all comes out in our camp sessions quite possibly this was the most important thing to come out of this one! I can hardly wait for camp. So I can come back and work with all these good people again in August! I left there about 7 AM Sunday morning. As I made my way along the east side of the Illinois River I thought of how my journey was coming to a close. I hoped Joe's car would get me the rest of the way through Illinois, Iowa, and into Minnesota without any problems! It did, when I got home the odometer showed I had traveled close to 1700 miles in all. When I got close to home I stopped in to check the field that I had left unplanted, it was dry! So when I got home around 2:00 PM I immediately hooked up the field cultivator and worked that field up, and of coarse 60% chance of rain was in the forecast for the next day! I got Joe up around 5 AM we had breakfast and got all the seed we would need loaded and was over in the field planting by 7:30 as we were trying to beat the rain! My dad stopped by around 9:30 Joe was planting and I needed to go get a little more fuel so I told dad I would buy him a cup of coffee if he came with. When I came to pay my bill the total came to $77.70 the cashier said that with three numbers like that I should buy a lottery ticket! I just thought I had the best three numbers on my side already God the father, Jesus Christ, and the Holy Spirit. With odds like that who needs the lottery! After dad left I noticed 7 seven pelicans fly overhead going south. Our weather was starting to head south also, because several times it was starting to sprinkle. We Finished and got moved back home. As soon as we got the grain drill cleaned out it started to rain. The wheat was planted another season has begun, and it was raining as I was writing some of this story! I am reminded of the words of the Apostle Paul.

> ***Acts 20:24:*** *However, I consider my life worth nothing to me, if only I may finish the race and complete the task the Lord Jesus has given me - the task of testifying to the gospel of God's grace.*

Gods' peace and abundant blessings to you all!

Your brother in Christ,

George Denn

Hey by George! June 4-12 2010

Luke 8:22-25: One day Jesus said to his disciples, "Let's go over to the other side of the lake." So they got into a boat and set out. As they sailed, he fell asleep. A squall came down on the lake, so that the boat was being swamped, and they were in great danger. The disciples went and woke him, saying, "Master, Master, we're going to drown!" He got up and rebuked the wind and the raging waters; the storm subsided, and all was calm. "Where is your faith?" He asked his disciples. In fear and amazement they asked one another, "Who is this? He commands even the winds and the water, and they obey him."

I have for a long time now known that Jesus can control the weather! Being God, and being its creator he seems to have a greater understanding about it than our modern weather forecasters! Who can be and a lot of the times are wrong. As a matter of fact some of the best times to cut hay, but not always is when, rain is predicted for the week! And some of the worst times, but not always, is to cut hay is when a long dry spell is predicted. But alas if one puts up hay as part of his lively hood sooner or later one loses a crop to the elements, and that is just the way it is! And when this does happen, it makes one feel that it has happened way to many times in one's lifetime! It does give me comfort though that we have a God we can communicate with in prayer to change our circumstances if he so chooses. I have a few story's one just recently, that could have sunk me financially if he hadn't done so! One of the first times that I experienced God intervening with the rain was right after he allowed 7 inches to fall on my first cutting of hay in June of 1996. By the time I could get it baled it was totally shot, as us haymakers call it! When it was time to put up the second crop of hay that year, it looked like rain was going to ruin that crop also, so predicted the weather man! I remember feeling the strain of it all! I also remember sitting at my kitchen table crying out to God in prayer about the hay. I just couldn't afford to lose this crop too, and that if it was his will that he protect the hay from being rained on. Amen! Later that afternoon my dad, my hired man Mike, and myself started to bale some hay. Almost immediately the rain clouds rolled in. My faith was being stretched pretty far as it was I saw a few drops of rain already starting to fall when Mike said, "Well George it looks like we are going to get rained out again." "Not this time Mike," I said. "I prayed about it today!" This amused Mike he laughed and said. "See it is raining right now; look at the drops on the baler!" And he pointed to them! My faith was just about gone. All of a sudden a great wind from the west blew up and blew those clouds, and rain out of there! As that happened Mikes jaw just dropped he didn't know what to say except that I had said it wouldn't rain at all. I laughed this time and told Mike that "no" I said that it wouldn't stop us from baling! And it never did! Shortly after this Ryan the other guy who was working for me at the time was cutting some more hay in another field came and commented. Man did you see that wind and he laughed! That evening for the first time I read these words from Psalms 104:1-7:

Praise the Lord, O my soul.
O Lord my God, you are very great;
You are clothed with splendor and majesty.
He wraps himself in light as with a garment;
He stretches out the heavens like a tent
And lays the beams of his upper chambers on their waters.
He makes the clouds his chariot
And rides on the wings of the wind.
He makes winds his messengers,
Flames of fire his servants.
He set the earth on its foundations;
It can never be moved.
You covered it with the deep as a garment;
The waters stood above the mountains.
But at your rebuke the waters fled,
at the sound of your thunder they took flight.

God showed me that he had showed himself in a mighty way that day, my faith had been strengthened!

Another time a year or so later, I was putting up some hay about a mile from here that I had rented from my neighbor Harvey. The hay was ready for bailing, but as I was raking it, it started to get cloudy. I felt a few sprinkles and prayed about the hay while I raked. My nephews were helping that day as well as some neighbor boys. It was raining all around us but where we were not a drop fell! When we went to unload we could see the rain on the blacktop road just up the road a little bit. Nobody said much. I did mention that it looked as if God were helping us out that day! The rain did hold off that day until we finished up when we pulled the last load onto the road it started to rain, but we got it unloaded before it got very wet!

This spring I started a new twist to my pumpkin raising! I planted some winter Rye last fall for a green manure crop. we had just gotten 1-1/2 inches of rain about the time I needed to work down the Rye. Rye is fairly tall so the ground was on the wet side when I worked it down. The soil turned lumpy as it dried, and my old shoe planter that I use for pumpkins doesn't like lumpy ground so it wouldn't cover the seed right. I tied a log chain behind so the seed would cover with dirt. This all left about 60% of the pumpkin seed lying in dry dirt! A week after planting only the seeds that got into the moister came up. These pumpkins are devoted to God so every year it seems that I experience some sort of struggle with them during the season! Once again at planting time I asked God if he would glorify himself through the pumpkin crop. I started asking God for 1 inch of rain. I also was asking others if they would pray for an inch of rain also. On Sunday the May 30th Rick Bengston from Fond Du Lac Wisconsin, called about 9 o'clock am, and said he was going to have his church pray for an inch of rain for my pumpkin field. There wasn't too much of a chance for rain and the low land where I put up my grass hay was dry so I decided I would cut it down. I kind of had it in mind that I would take a few days off next week, and go up north and visit my friend Gordy near Duluth, and Tom up in Orr, and get my canoe that I had purchased from a guy we call Bill on the hill! Tom had planned to bring it last fall when he came down to help pick pumpkins ,but something went on at the last moment and he wasn't able to bring it. Joe would be gone for a youth event through his church, so If we could get this work out of the way it would give me some time between haying, and working on weeding the pumpkins.

About 5 PM it started to rain, and I finished cutting the field down below the house in the rain! It had rained straight down until 6:30. I had gotten an Inch here but you didn't have to travel far in either direction and they only got a couple tenths! My pumpkins will now sprout because it rained.

> ***Matthew 5:45:*** *He causes his sun to rise on the evil and the good, and sends rain on the righteous and the unrighteous.*

Well here it is almost 2 weeks later and it just keeps raining! I did make the trip up to Orr and back not because I had all my work done, but because I couldn't do any more until it dries up! I went first up to Gordy's for a couple of days. As we were out fishing Gordy noticed his brother in law Carroll was by his boat so we went in and invited him to fish with us. Gordy likes to get under Carroll's skin which is amusing if you know them! Carroll likes to keep blue gills and Gordy thinks that they are not worth keeping so every time Gordy would throw one back it seemed to pain Carroll. So I told Carroll that I wouldn't aggravate him like Gordy does! Before we went to pick up Carroll, Gordy had asked if I ever heard anything from Ray and Denise Olson. I mentioned that I had visited with them last at the Wisconsin Dells last Sept. at the festival for our church down there. I had left my cell phone back at my pickup to charge it and I found it interesting that I had a message from Ray Olson on my voice mail. I finally did get in touch with Ray when I got back home He had a pastor friend that was going to be in Mankato shortly and was wondering if I knew any churches his friend could speak at? This man was from India was a Hindu that turned Christian he seemed to have an interesting story. So I told Troy and Jeff about him I hope they get connected as he sounds to have an interesting story! On Tuesday Morning I said Goodbye to Gordy and Bonnie and Drove up to Orr. First I drove out to Mitch's to get my canoe. Mitch is Tom's son. After the canoe was loaded I drove over to Tom's place to wait until he got off of work. Tom came and we ate lunch. We had planned on going fishing that afternoon, but because it was raining we just sat around and talked for a while. Around mid Afternoon we decided we would go out to see if Bill on the hill was at home. He was and we spent the next couple of hours talking with Bill. Bill lives on top of a hill in a new house that reminds me of a chicken house. Bill calls it half a house! It seems to be in the tree tops, but Bill has quite a view from there! Bill was talking and all of a sudden he was telling us that he had an experience a couple of weeks back, but didn't know if he should tell us or not. Evidently he was in his house in the daytime when all of a sudden this incredible bright light started shining in. Bill said the light was so intense and so pure that he just knew it was God. And that his kingdom was near if not here already! I mentioned to Bill that I had a lawyer stop out to my place one time looking for some hay to cover some cement during some cold weather. He mentioned he was a Christian and was telling of an experience with a bright light that he had. On night he woke and this bright light was hovering over his bed he woke his wife up to show her. He too felt that this was a visit from the Lord! Tom mentioned that he also had two experiences with a bright light! Once after a Bike accident, and once just before he became a pastor. Evidently this bright light showed up in his room and sort of took him a back the way Tom tells the story this bright light entered his body where his heart is! While I was Listening to these story's I was thinking I wish I had my own "Light" experience! But wait a minute I had. When I was 15 years old one evening about this time of year. I had went to bed my window was open as it was a hot night. All of a sudden this great wind blew so hard it Blew my tobacco can out into the hallway off of the nightstand! I heard what sounded like log chains fall to the floor and my room filled with a light so bright that I could see it with my eyes closed. I never prayed so hard in my life as well as a 15 year old Catholic boy could pray! I always felt that there was some kind of evil connected with this presence, because of the sound of those chains. As I was sitting there telling this story it came to me that I was visited by God that evening and the sound of the chains rattling was because Jesus was freeing me from the chains that had me bound. Before I realized it or cared! So I could follow him at the proper time! Wow! What a revelation!

John 8: 34-36: *Jesus replied, "I tell you the truth, everyone who sins is a slave to sin. Now a slave has no permanent* *place in the family, but a son belongs to it forever. So if the Son sets you free, you will be free indeed.*

I had heard the chains drop even if I couldn't see them! That was something, and it took me 33 years before that episode became clear to me!

I know now that this trip was no accident and on my way home I had a couple of interesting happenings. One was I wanted to check my gas mileage so I stopped at the gas station which is called Freedom right before you turn onto hwy 33 off 53. After I filled my tank and found out the mileage it was raining from Orr to there, but immediately when I turned onto hwy 33 the sun came out and the rain stopped! When I stopped at my house I found it was exactly 301 miles and no tenths from Tom's house to mine. Usually this drive takes me 6 hours + and usually I drive faster than the speed limit. It took me 5 hours and 22 minutes this time and I drove exactly the speed limit all the way! I got home Checked my fields the Rain has pretty much ruined my 70 acres of hay that I cut before my trip, my pumpkins need cultivating but the rain has put a stop to that also! This kind of stuff used to bother me a lot! But my Identity is in Christ not what I do! And I also know God is fully able to Change our circumstances when he wants as you have just read.

Psalm 9:1-2: *I will praise you, O Lord, with all my heart; I will tell of your wonders. I will be glad and rejoice in you; I will sing praise to your name, O Most High.*

Gods' peace and abundant blessings to you all!

Your brother in Christ

George Denn

Hey By George! August 11, 2010

Psalm 115: 1- 11: *Not to us, O Lord not to us but to your name be the glory, because of your love and faithfulness. Why do the nations say, "Where is there God?" Our God is in heaven; he does whatever pleases him. But their idols are silver and gold, made by the hands of men. They have mouths, but cannot speak, Eyes, but they cannot see; they have ears, but cannot hear, noses, but they cannot smell; they have hands, but cannot feel, feet, but they cannot walk; nor can they utter a sound with their throats. Those who make them will be like them and so will all who trust in them. O house of Israel, trust in the Lord - he is their help and shield. O house of Aaron, trust in the Lord - he is their help and shield. You, who fear him, trust in the Lord - he is their help and shield.*

Here it is August already! Both summer camps are over and it's about three weeks until it's time to pick pumpkins again! For those of us who always wonder, just where has the summer gone too? I will attempt to remember where mine has gone. Ever since the 30Th of May it has been extremely wet here. I had 76 acres of first crop hay that never did get baled it just laid there after I had cut it, and the second crop grew up through it! All field work had been delayed many days! The only way we could keep the weeds at bay in the pumpkin fields was to get a few more guys to help, and go through walking with back pack sprayers. We did 60 acres of pumpkins this way and 4 acres of corn. Needless to say that was quite a feat! On days that I could I was riding my old WD Allis Chalmers cultivating. This tractor is about 12 years older than I am so it has much to be desired as far as creature comfort!

One day after lunch, on a day when I could cultivate. I had so much going on in my head, and trying to make up for lost time. I did something really stupid! This tractor has an alternator on it that needs to be polarized to charge the battery. After the tractor starts you have to take a piece of electric wire up by the alternator to do this. So instead of getting off and then back on again I just have been starting it from the side while standing there. This time I started it, but as the engine came to life I realized I had left it in gear! Just as it was coming forward the tractor hesitated long enough for me to grab the clutch pedal and I was able to take it out of gear, and stop it's forward motion, before it ran me over! I will always believe it was God himself who made that tractor hesitate, If it hadn't, it would have run me over and I wouldn't be writing this right now! The very next day I was driving my pickup about 5 miles from here. I was flexing my fingers as I was driving. I am starting to feel some arthritis or something like it! As I was doing this I was thinking of something else other than my driving. Suddenly I realized I was driving off the road! I did get my vehicle back on the road. The ditches were very steep right in this place, so once again I thank God for his help! This year Northern light camp was held on July 11-17, but I had to be there on the 10th. I had to get my second crop of alfalfa harvested before I went, which we did. Also we had to get some wheat bundles cut, but because of all the rain we had been getting this was somewhat harder! The old binder didn't skip a beat which was a God thing in itself! I had to take a pretty small cut so the drive wheel wouldn't slide and after 6pm the mosquitoes were unbearable but other than that we finished the day before I left. When I got to camp I was so tired that I knew I just wasn't going to make it through camp without some divine intervention! So after the days training session I asked Todd Fox and Troy Meisner to anoint me with oil and prayer. After they did that, almost immediately my tiredness left! I was the councilor for the boy's dorm 12-14 year olds. Michael Haack was co-councilor. One night our dorm had been talking about something. I can't remember just what we were talking about. One of the boys said. "You have to be careful, because things are not always what you think they are. One time I was acting up in church and my dad said lets go into the play room. I thought my dad was going to play a game with me, but I got a spanking instead!" My apologies to this young man, but that statement just cracks me up every time I think of it! One of the activities at the camp was a pottery session. Taught by Rick Bengtson I made a coffee cup that I actually can drink coffee from. There were three Hi ropes courses I only did the easiest one. I'm still not too good with heights! They had a rock climbing wall this year I actually did some climbing but only got a little over half way up. One of the guys from my dorm did the best at rock climbing. We went river canoeing down the Root River that flows around the camp. I had Hunter and Josh in my canoe. They were such good paddlers that I hardly had to do anything but steer, and we were the first ones in our group to make it to the landing. On Friday evening they had a square dance it was way more fun than I remember from high school. Friday evening was also the time for our dorm photo, so I had my bib overalls on when we had our picture taken. One night they had a movie playing. My dorm had the option of staying in the dorm or watching the movie 'Twister'. They chose the movie. Half way through the movie I was getting sleepy so I told Michael to bring the boys back to the dorm when the movie was over as I was going to go to bed. After I had left the discovery center the place where the movie was playing. I found that the doors were all locked to get into the buildings were the dorm was, and remembered that the staff lock up everything around 10 o'clock. So that meant I couldn't get back in where they were showing the movie either! I tried calling several people, but like me they either had their phones off or had them on silent mode! What a place I thought to myself! I just had to wait until the movie was over and enjoy the mosquitoes. While I was waiting I heard a bunch of coyotes howling down below the camp. Did you know that if enough start howling at once it sounds like a police siren? As I sat on the front steps a mama raccoon and 3 young ones came walking up the side walk I gave a small shout and the 3 young ones climbed up a tree! Finally the movie was over and I was able to go in with the others and go to bed. The messages for camp as well as for the Heartland camp were about our inclusion in Christ.

> ***Ephesians 2:10:*** *For we are God's workmanship, created in Christ Jesus to do good works which God prepared in advance for us to do.*

To understand the depth of that statement and to put that mentality into practice in our everyday existence seems to be a lifelong process! The hardest part of camp is when it's all over and it's time to say goodbye to every one! At least for me it is. Rick had me take a couple of his potters wheels home with me because he didn't have the room for them in his car, and I would be going down to Heartland camp in Illinois in a couple of weeks anyway. When I got home I had to put the wheels in my house so they would be safe from the elements. As I was unloading the wheels one of the chickens darted into the house and I had to chase her out of my living room! I was teasing Rick that I hoped he didn't find any chicken stuff on his wheels. I put on my overalls and went right to work. Joe and David had got done with the spraying so we put the wheat sheaves in the shed as they were predicting rain for that evening. Joe had already cut the red clover hay while I was at camp so I started cutting grass hay the next day. we also had to get 3rd crop alfalfa up again before I left all that adds up to 170 acres of hay we put up in 2 weeks without any rain! I swathed the 50+ acres of wheat hoping we could get that done too, but it was not meant to be! While we were putting up the hay I saw rain clouds go to the south of me, and to the north of me, so we saw God actively helping with the hay, but when it came to the wheat the more I prayed about it the more it seemed to want to rain! Sometimes I really wonder about God what he is trying to get through to me. So I had to leave the wheat to Joe and neighbor David to contend with! 50 acres of wheat could be worth up to around $15000 this year but I had to give all of that over to God, because I would just let it bother me if I didn't. I don't generally call home while I'm at camp because usually all they have to tell you is bad news anyway so instead of stewing about things back home, I am generally oblivious to the outside world while I'm at camp. Ignorance is bliss! I picked 8 ripe pumpkins before I went. I wanted to give some of them to my cousin Nancy when I went to her place after camp. I drove as far as Iowa to the first wayside rest and took a nap for an hour before I continued on. Boy was I tired! after a couple of days at camp, a couple of the staff in training people carved 2 of the pumpkins and I saved the seeds and counted them out 1010 seeds were in those two pumpkins. They had a contest between the dorms who could guess the closest to the actual number would win $20 to spend at the camp store. I forgot who won, but I roasted the seeds the night of the dance and ate most of them during a card game. I think that's the first time I played cards during a dance! And I did give some of the seeds for others to try. It really has been fun working at these camps the past 8 years, and to be involved from the beginning of them. Too watch them grow, and see some of the youth that once were campers become staff. It's hard to explain exactly what that is like, but it really is a wonderful thing! I had a chance to donate to several who were experiencing financial difficulty. I had plenty of funds with me, and Jesus tells us in Matthew 10:8 freely you have received, freely give. It sure does feel good to be able to give for a change! It's a funny thing though the next morning I was counting the money that I had, and there was $30 more in my billfold than before I gave the gift, and my billfold was locked in my pickup, and no money exchanged hands during the camp! The first night I lost my reading glasses, and I was also a roommate to a person who snored all night. I prayed about these things and Steve gave me his extra pair of reading glasses and a set of ear plugs so the person that snored all night was no longer an issue for me! For some reason stories from our past generally get told at camp. This year I remembered 2 that I revealed the first one was how I taught my sister Jane to drive a 4 speed. My dad had a one ton 1959 Chevy truck when we were all home. I was supposed to teach my sister how to drive it. So I took her down to the hay field below our house. After she had made a couple of laps I assured her that she was doing fine and I jumped out of the truck! I still laugh at that although it probably wasn't too funny for her at the time! The next story was how I taught one of my sister Jane's girlfriends how to ride a horse. I helped her onto the saddle and handed her the reins then I went back and slapped the horse on the rump and hollered get up! As I recall she screamed and I told her to hang on tight, I was laughing fit to kill! Years later I was at a street dance and that girl came and gave me a big hug, and told me she never was so scared in all her life! There is also the smoke bomb story, but it burned up there sleeping bags, and I really got it for that one, so I will leave that one for another time! See how much I have changed since Jesus is Lord of my life! The week at Heartland camp sure went quick. Once again saying goodbye was the hardest part of camp. I hope to do it all again next year God willing! I drove east on hwy 17 to my

Cousin Nancy's place. I was stopped at a red light in Kankakee and I fell asleep. When I woke with a jerk I said to myself this is no place to be doing that! I did stay awake those last 18 miles. I gave Nancy those 6 pumpkins that I had bought to camp. I had a good supper and went to sleep from 8:30 pm until 6:30 am. Nancy's mother Donna rode back with me to Minnesota the next day. We got back to my dad's place around 4 pm. I left around 5 to see if any wheat had been combined? They had got one field combined and a start on the second one but once again rain had stopped them. We were able to get the straw baled off that one field before it rained again. So at the time of this writing half of my wheat is still lying in a windrow out in the field for exactly 3 weeks. I find it is easy to have faith when you see God at work all around you, but very hard when you don't see him and your lively hood is at stake!

> ***Job 13:5:*** *though he slay me, yet will I hope in him.*

So this is something I must trust God for, as he stretches my faith in him. And even if my wheat is lost I know God will work something out further on he always has!

> ***Psalm 118:16-17:*** *The Lords right hand is lifted high; the Lords right hand has done mighty things! I will not die but live, and will proclaim what the Lord has done.*

Gods' peace and abundant blessings to you all!

Your brother in Christ,

George Denn

Hey by George! Oct. 6, 2010

Mark 10:28-31: *Then Peter began to say to him, "See, we have left all and followed You." So Jesus answered and said, "Assuredly, I say to you, there is no one who has left house or brothers or sisters or father or mother or wife or children or lands, for my sake and the gospel's, who shall not receive a hundredfold now in this time- houses and brothers and sisters and mothers and children and lands, with persecutions- and in the age to come, eternal life. But many who are first will be last, and the last first."*

I had pretty much made up my mind that I wasn't going to go to the Wisconsin Dells fall celebration this year. My church Grace Communion International Sponsors this event. I had just received 6.5 inches of rain a week earlier and I was just about to e-mail Betty Johannsen that I wouldn't be able to bring the pumpkins, squash, and gourds that she wanted me to bring to decorate the worship hall with. Also that weekend of the event an antique car motorcade was supposed to come to my place and purchase up to possibly $1000 of worth of merchandise from my stands. So said the lady that I donated $16 worth of squash to for a door prize! She said I just had to be there so people could give me the money instead of just putting it in the pay box that is down there! That very evening while Joe and I were on the run, as we call it to collect the day's money, I was telling Joe all the reasons I wasn't going to go to the Dells. I went to the mail box first before I checked the box for money. There was a letter in there from a guy named James that I had met at camp last summer. I opened that and inside was a note with a check for $500 inside! The note said something like have fun at the Dells! I was thinking If someone went to all that trouble maybe I should reconsider! I thought I should call Doug Johannsen the Dells coordinator to see what time everything started down there, but before I could call him he called me. He was asking about all the flooding that was going on around St. Peter. In the course of our conversation I asked him about when everything started down at the Dells. I already had

room arrangements back when I was conside
thinking that I should call my sister Jane an
ment to get my hair cut, but before I could ca
wondering if I could bring her some pumpki
so I asked her if it would work to cut my ha
"Yes I can do that", she said. The very next day Doug Joh
called me again. Doug told me that he had arranged an interview with Mike Feazell vice president of Grace Communion International about my books. Something was telling me that this wasn't going to be just an ordinary weekend! Another person sent me $300 dollars for the Dells. I felt God really wanted me to go to this because money wasn't the issue it was time! I also needed a new bible my old ones were shot! I had bits and pieces of them lying all over the house. I needed to get some pumpkins and squash washed up for Betty Johannsen and a load of pumpkins for Troy Meisner for his Spring Valley stand. Also typically this is the busiest weekend of the whole year. So it was real hard in every way to break away from here! About 5 o'clock the evening before I was to leave Joe's brother David called that he was stranded in town one of the pickups were broke down. Oh yes! Just what I need! We did manage to get the stuff loaded and washed that I was taking the next day but I was too tired to go to town that night for the bible and clothes that I would need! So I just put my swimming trunks in my suit case, and resolved to buy the clothes and bible I would need the next evening when I got down there! Terry and I were pulling the pickup that had broken down to the guy that repairs my vehicles when Tim Strommer called. Tim was telling me that he had a new bible for me, and that he would even deliver it! Tim showed up Just as I was about ready to leave he also had a cup of coffee to go for me! After giving final instructions to my workers about covering the stuff as it was supposed to freeze this weekend I said goodbye and I was out of here! The time was 8:30 am an hour and a half later than I intended! I had to stop off to my dad's house. I was half way there when I realized that I had forgotten to bring some copies of my books for the interview. I wasn't going to go back home now my dad had a few copies so I would just borrow those. Why does it always feel like I am escaping

m something when I go to these things? I got down to Troy's stand in Spring Valley unhooked my trailer and continued on my journey. I was staying in the same room with Troy and his son Ian and I learned that they would join me sometime the next evening. I got to the Kalahari resort about 3 o'clock in the afternoon, only a half hour late for the meet and greet session or in other wards food! I no more than stepped one foot into the doorway when I hear someone holler George! I looked around but could see no one that I knew? About halfway down the hall someone ran up beside me and said George I hollered at you but you didn't wait! I could hardly believe my eyes it was Ky Swamp! Ky stayed at my place from the 1st of April until about the 5th of November in 2005. I was startled to see him and he looked a lot different than he had the last time I saw him about 3 years ago. I scarcely knew what to say to Ky with all the mayhem that went on to get there and him looking so different basically I was speechless. I got a chance to talk with Ky again after services my thoughts were a little more collected then. I just couldn't get over how much he had changed! In the course of our conversation I invited Ky out for a meal sometime that weekend. He said that would be ok, but it never came about. After a small party at Floyd and Carolyn's room I went shopping at Wal-Mart's for some clothes and reading glasses because I can't read a thing anymore without them. I even showed Rick Bengtson that I wasn't kidding that I only had my swimming trunks in my suitcase! After my $100 shopping spree I went back to my room and crashed for the evening! How nice it was just to go to sleep without thinking of what I had to do the next day, because services didn't start until 10 am the next morning! On Friday after services I did see Ky again, but it wouldn't work out for him to go get something to eat that day. So with nothing to do and no plans until 2:30 I went to the water park and sat in the hot tubs I felt like I had died and went to heaven! I also floated down the lazy river a couple of times. Then it was time for the breakout sessions I went to the one Willard High was giving but I forgot what his topic was! I was heading to a small party that Mary Campbell was throwing, but on my way I ran into Josh Zuniga and he mentioned that several people were going bowling around 9 o'clock that evening. But that was not to be ether when I got to the bowling alley no one was there. I was getting tired anyway so I just went back to my room. I had come here to relax any way so it didn't bother me much not to be doing anything. Shortly after I got back to my room Troy and Ian showed up so I talked to them until I went to sleep. Saturday after Church services Isaac Bryant and I was supposed to meet Doug Johannsen. Doug took us up to Michael Feazell's room for our interview. We were first introduced to Nathan Smith; he's the video and light guy for GCI. After a bit Michael appeared and the interview proceeded about my books, my involvement in youth ministry and highlights of my time working in the youth camps in which I serve. Michael asked me to read one of my stories. Since Isaac was next to be interviewed I read about a story I wrote of Isaac when he was one of the youth that were in my dorm when I was the councilor. While Issac was being interviewed I snuck around and listened to what he had to say. A couple times I was almost in tears I hadn't known God had used me so much to influence Isaac's life! After Isaac had finished his interview I opened up the door and revealed that I had been standing there! We all had a good Laugh! After the interview I was invited to Tracy and Cheri Porter's room for lunch the Porters always have the most entertaining people at their place and entertaining talk going on! Tracy had to hurry because he had to do one of the breakout sessions. The general Theme for this year's celebration was 'on to maturity'. Christian maturity that is, and how that should apply to us. After the last break out session that Michael Feazell gave, Doug Johannsen invited me along to dinner with he and Betty, Michael Feazell, Nathan Smith, and Sam Butler. It was a great evening! I had a great time talking with Nathan and Michael and the others. As always I Felt blessed to be amongst this group of people! After praise and worship a pizza party was taking place at the new theme park for the youth and those who work with them. After a round of mini golf and a ride on the Ferris wheel I decided that I had enough for one evening. Remember I came here to relax not to do everything there is to do! So off to my room to sleep! Troy and Ian was there again so I talked with them for a bit before I went to sleep. On Sunday morning I packed up my things so that when we went to breakfast I would be out of my room. So I just hung around the worship hall talking with friends until services started. Sam Butler Gave the closing message from 1 John how Jesus is love. After services there was nothing more to do but to say goodbye to people I knew and those I had met. I had walked out of the worship hall and was talking with James Shunkwiler and Brenda Hill. Ky swamp walked up he said I just wanted to say goodbye to you George, and he extended his hand. I looked at Ky and said "we can do better than that can't we Ky?" I gave Ky a big hug and told him how proud I was of the fine young man that he has become! Ky said at the very same time we're buddy's George, we're buddies! I was crying when I said that to KY I could tell he was holding back his tears also! Ky was 15 the year he was here he is 20 now. For me that moment was the highlight of the celebration! Rick Bengtson took some pictures of Ky and me so I could show the folks back home. Ky was saying to me that there had been a lot of bad things that went on the year he was here. I told Ky that I never thought of those things just the good times that we had, and there were many. I also told Ky that I wouldn't trade that year he was here for anything in the world! Once again God showed me how he had used me to influence a young man! I was showing Tracy Porter a picture of me and Ky. Tracy said its interesting George that the Word Ky in the Hebrew language means 'just like' Ky is just like you George! Tracy stated. Well I hope Ky is better than me believe me it wouldn't take much!

After I returned home I found out the weekend had went fine. Everything had got covered that needed to be covered so the frost didn't hurt anything! So often I think of the statement Dan Rogers made at a conference. That sometimes you must leave the work to the hired men just like James and John did when Jesus called them!

Mark 1:19-20 When He had gone a little farther from there, He saw James the son of Zebedee, and John his brother, who also were in the boat mending their nets. And immediately He called them, and they left their father Zebedee in the boat with the hired servants, and went after Him.

I figure that is why God put them into your life in the first place! And the antique car thing was pretty much a joke I guess only 8 cars showed up, and Terry told me that one person bought $2 worth of squash! So it would have been a real travesty if I would have stayed home because of that!

> ***Ecclesiastes 7:12:*** *for wisdom is a defense as money is a defense, But the excellence of knowledge is that wisdom gives life to those who have it.*

Gods' peace and abundant blessings to you all!

Your brother in Christ,

George Denn

Hey by George! Oct. 26, 2010

Zechariah 4:6: *'Not by might nor by power, but by my spirit,' says the lord of hosts.*

I was asked if I would write a story about my pumpkin farm. I thought to myself there has been so much that has gone on since I started raising them where do I begin? I guess the beginning is a good place to start! I never had any aspiration to write nor did I have to raise pumpkins, but here I am doing both! Since how I began to write is another story in itself. I will just concentrate on the pumpkins that I grow. I was struggling here on the farm that I was born and raised on. As the 1980's-1990's came and went. The make-up of the family farms as I knew them were changing rapidly as well. I myself internally was struggling as well! As I slowly left my former way of life to enter into the life God had marked out for me. This journey lead me to the World wide Church of God on June 17 1995 into a Church that was in turmoil as well! Since I was a farmer all my life I didn't particularly want to quit and do something else just because others had done that. There never seemed to be an open door so to speak. I asked God if it was his will that he would give me something right here to serve him with and that I could make a living on. First he gave me a business raising and selling hay and a slogan 'Hay by George!' After that an Idea started to form. If I would cut some corn stalks and put them into bundles I thought people would buy them for fall decorating. I purchased an old corn binder a machine that was used in the early part of the last century for cutting and tying corn into bundles. I figured if my idea was a hit I would somehow have to mass produce these, and this antiquated machine was the only way to do it! I sold 9 corn bundles that 1st year for the sum of $45 I did this after the 15th of Oct. After a couple of years I added straw bales for sale with the corn bundles, and I found out that by taking them to the end of my driveway on an old truck my sales were much better! One day by chance I was in town at a lumber yard when I ran into an old friend of mine from my former life, Tony Foty. In the course of our conversation I found out that Tony was raising pumpkins. I told Tony that I may be in touch with him the next fall as I thought pumpkins may go well with the straw and corn stalks! I did call Tony the next fall all he had left was 41 pumpkins left that the stores he sold to would not take! Tony bought them out and in two days they were all gone! I bought them for $1 apiece, and sold them for $2 each! Tony and I kept in touch after that and by springtime Tony reviled to me that his partner in the pumpkins was eliminating him from the equation! I was telling Tony that I had this whole farm that I guess would grow pumpkins and that if he didn't mind me being his partner I wouldn't mind trying to raise them. The first year in 2000 we grew 4 acres of pumpkins. Through working together that first year I was able to share my faith with Tony. He saw first-hand how Jesus Christ had change my life. Tony asked me if I would baptize him. I got my minister at the time Charles Holladay involved to see if he felt Tony was ready for that. He felt he was. So on July 4th 2000 I baptized my friend Tony. Tony and I dissolved our partnership after our second harvest season. The acreage that was planted to pumpkins steadily grew from the first year of 4 acres to the 60 acres I grow today. I have no intentions to grow more acres, because 60 acres of pumpkins is a lot of work, and I am no spring chicken anymore! From the beginning I choose to use the honor system. Most people can't believe that I run the business that I do through the honor system! As this year comes to a close it looks to be my best year yet! The high lights of all this pumpkin raising are all those people that have worked with me over the years, and those that I have met because of it. A comment was made by my friend pastor Doug Johannsen "George you sure spread a lot of joy with these pumpkins." Since that time I tell everyone that is helping here that they are involved in the ministry of joy! So I guess this is my ministry or better yet Gods ministry that he has put me in charge of! Another aspect that has transpired over the years here is what I call 'the pumpkin thing' I called it that for lack of something better to call it! This is an annual fund raiser for the winter camp 'Snow Blast'. A three Day event that takes place during mid-January near Rochester Minnesota that our district of Grace Communion International sponsors. Even that event that is in its 6th year now has grown to the point that we have to hold it on two separate weekends. This past year it was held on the weekend of Sept. 11 and Oct. 10th. People that are most generally associated with our fellowship show up. Some driving as far as 6 hours enjoy the weekend. Everyone somehow miraculously fits into some sort of role whether to help with the food pick pumpkins drive pickups watch the real young ones or keep my house clean. My biggest challenge seems to be everywhere at once which of course is impossible! This Year I had three people that showed up from six hours away. They cleaned my windows as well as fixed the window panes that were broken! I was wondering if I would ever get around to doing that! One of the Guys was from Orr MN. he came back a week later and helped with the rest of the pumpkin harvest as well as help us to put a new roof on my house. We were able to raise $2500 for snow blast this year. It is more than just picking pumpkins I think fellow shipping is the biggest thing that goes on at that time. My neighbor Jon bought his horses over and gave a couple wagon rides and Rick Bengtson came from Fon Du Lac Wis. with his potters wheels so there was some pottery being made during the evenings. It was because of working with the youth at several camps our church sponsors that I have come to realize that God is using this pumpkin business for things greater than what can be seen! I always make the comment that "you cannot send a youth to camp without it changing them! You cannot go to camp without it changing you!" I say that from my own personal experience!

It does take a lot of effort to raise 60 acres of pumpkins. I was talking with a lady at the Wisconsin Dells celebration I was telling her that I had to make the bulk of my yearly income in 61 days! She was asking what I did the rest of the year? I told her I get ready for those 61 days! There are a lot of uncertainties in my own life at this time. I rent this farm from the same family that my father did and the guy that owns it is in his 90's. So I have been here about 49 years. The people who help me here come and go from year to year. So it really makes me rely on God for the answers for it all!

Like a spreading pumpkin vine you never really know where it's all going to end up! Every year something interesting always ends up in the pay boxes. Just the other night I got this note in the pay box at my Eagle Lake sight I assume it from a young mother it reads as follows. "I am truly sorry I do not have the money to pay for two pumpkins. I have two little girls that want one, and they have had a very hard life. I just want to see a smile on their faces and if a pumpkin can do that I will pay in any way that I can in the future. I am sorry I do not have the money to pay for them I am not a thief I would not normally do this cuz if I had the money I would pay."

That was probably the most touching letter I ever received in the boxes. I'm glad she took the pumpkins I always have said if people were in need I hope they just help themselves! I also prayed for this person that God would bless them and the 2 girls. It's because of letter like that one, and to see probably in the thousands now family's taking pictures amongst the pumpkins, and to see the smiles on children and the not so young as well that keeps me doing all of this. I can really see our triune God at work in all of this! And I know for the time being that God has me exactly where he wants me to be. I guess until he moves me somewhere else this is where I will stay. You can read about many more of my adventures with the pumpkins and the youth camps where I serve, and the people I have met in my two books Hey by George! And Hey by George! II.

> ***John 15:5:*** *I am the vine, you are the branches. He who abides in Me and I in him bears much fruit; for without Me you can do nothing.*

So to God be the glory!

Gods' peace and abundant blessings to you all!

Your brother in Christ,

George Denn

Hey by George! November 30, 2010

Jeremiah 30:19 *then out of them shall proceed thanksgiving and the voice of those who make merry; I will multiply them, and they shall not diminish; I will also glorify them, and they shall not be small.*

Ahh! Thanksgiving, one of my favorite holidays. A few weeks ago my friend Gordy asked me if I had any plans for Thanksgiving. I said, "Not really." Gordy said that I do now! He asked me to spend the Holiday with him and his family, and if could come up with something to speak to his congregation in Duluth, on Saturday, that would be alright too! Now the thought of Thanksgiving dinner is a great one, but the thought of coming up with something to speak on sort of makes me want to panic! I feel I am rusty in this area and would rather not, but sometimes you have to do what you don't want to do! What I decided on was to read them the last story I wrote of my experience down at the Wisconsin dells. There I made it easy on myself! My plan was to drive up to Gordy's sometime on Wednesday the 25th. Once again I found myself with only $40 to go on the 230 mile trip. Make that 330 miles, because I told Pastor Tom up in Orr that as long as I was up to Gordy's I might just as well come up to see him also. Tom was saying that as long as I was coming that maybe I could speak at his church on Sunday as well! Sure why not, as long as I already had something picked out I would use it at both churches! At about 8:30 in the evening I called my friend Wayne. He owes me some money on some hay that he bought from me earlier in the year. I figured if he bought up the subject of money that we would talk about it if not we wouldn't, because I don't want our relationship to be just about money. Wayne is in the process of selling his milk cows, so after this year it looks like I will need a new hay customer! It's all kind of a traumatic process for him. I know I went through the process of selling my cows 16 years ago I'll never forget the day I called the auctioneer: after I made the phone call I went into the living room and I bawled! I reassured Wayne that Things will be fine he will find something else to do. God wouldn't let him starve. I mean look at me I weighed 160 pounds 16 years ago, now

I weigh 230 pounds. I ask, does that look like starvation to you. I should say not! Money didn't come up in our conversation, so after our chat I started packing my suit case wondering where the funds were going to come from for my trip. Just then my phone started ringing it was Wayne again. He was asking me if I needed any money for my trip. He said a thought ran through his head that maybe I needed some money for my trip. Yea I said I only have $40. Wayne was saying that he didn't want to mess with my faith and all, because he has heard of the story's where I had left with as much! But if I could use some funds he had some so I asked him what he could spare and he thought about $1000. This would pay what I owed my hired guys yet and leave me $200 to go on! So I told Wayne Just to pay Joe and Joe gave me $200 to go on. At 5:30 am Terry Called me and woke me up he told me I ought to get going early as there was a huge storm moving in. I thought it would be nice to miss that if possible! I took the best vehicle I own a 92 ford half ton pickup, it only had one worn tire and the spare wasn't much either, but it has all the lights that work and a heater; sort of. Plus the windshield wipers that work if that sounds wild you ought to see the other two I have! First I stopped off to my Dad's in Waseca, but I left earlier than I had planned, because of the storm. On my way David Maki called he was telling me that he was basically snowed in at his cabin and was wondering if he could get out could he stay the winter down at my place. Sure I said I told him that when he left after the work was done last fall. He was telling me that his place was six miles from any road David's place is about an hour drive north east of Orr MN. Latter I did call Tom Kennebeck to see if he had any Ideas on how to get David plowed out. Tom did mention that he had a friend that lived near there with a snowplow. David did get himself out I thought it was kind of funny David was coming down to my place while I was Driving up to the Town he lived near! I heard of this on Thursday and Friday I was thankful David got out; but back to my journey up north! I got about half way it was about noon anyway when I stopped at a McDonalds, but when I came back out to my pickup it wouldn't start. I just hate that; why does this always seem to happen when you're in the middle of nowhere? I said a prayer for my lunch and my vehicle. First I ate my lunch and then I put on my coat and gloves and went to see what I could see about my vehicle. I opened the hood and pounded on the battery cables, and I tried jumpstarting it on the solenoid! After this procedure I tried the key, nothing! I was able to move the battery cables ever so slightly. I tried the starter again this time the engine started, and was able to proceed on my journey; I said thank you Jesus as I drove away! I didn't shut it off again until I got to Gordy's. When I stopped for gas I just let it run, I figured if somebody wanted to steal it they deserved the thing! I made it to Gordy's about 2 in the afternoon. About Dark it was really snowing heavy so I was thankful to be safe and sound for the holiday! Gordy was telling me that this guy by the name of Mark had been reading my stories, and he said he would like to meet me. Mark had sent Gordy this newspaper clipping of two guys in Wisconsin that were raising a certain kind of pumpkin for its seeds. They pressed the seeds to get the oil out of them. I was especially interested in the machine that they used to de-seed them with! I had been pondering about how to remove them in an easier way than by hand. So I will have to look into it, because I figure there is no need to reinvent the wheel if a machine already exists. Gordy had set up a meeting with Mark on Friday, but first things first. Thanksgiving Day Bonnie had Turkey with all the fixings, but the best of all was the smoked turkey! Mm, mmm! I ate so much that I felt like if I had so much as a grape more I would have exploded! On Friday we went to see a lady from Gordy's church. She was in a nursing home being re-habilitated. She didn't seem too happy to be there I guess I can't blame her for that. I don't get too excited about nursing homes myself! Then we went to meet Mark and his wife Darlene for lunch. Darlene is a member of Gordy's church; Mark is catholic, and a philosopher. Mark was telling us that he grew up on a farm in Nebraska, so he could well relate to my stories I wrote. We talked about many things as we ate, but just before we parted way Mark shared this story with us that he has been pondering about since 1943! He and his brother were back home thinning a field of carrots. The rows were long so they would both start on the end and work toward each other. Mark was eating some of the carrots as he went on his hands and knees down the row. Mark started to choke on a carrot; he was patting his chest trying to get his brothers attention, but his brother just mimicked his gestures! He tried patting his back; once again his brother just mimicked him! Suddenly Mark went unconscious. All of a sudden Mark felt a strong hand hit him in the back it dislodged the carrot; he felt the carrot hit the roof of his mouth and he spit it out. Mark was able to start breathing normal again! Mark always felt that it was his brother that rescued him; and so gave him the credit. Once upon hearing the story his brother got agitated and said to Mark that you always say I helped you that day, but I didn't I was nowhere near you at the time, so quit saying it was me that saved you because it wasn't me! Mark was asking me and Gordy if we thought that was a light at the end of the tunnel story. I told Mark that I firmly believe that that was God that helped him that day! I really enjoyed our talk with Mark I hope we can talk again someday. Friday evening was left over day. Everyone just came back that was there yesterday to enjoy the leftovers! So I could have some more of that delicious smoked turkey! Saturday morning we had church so I just rode with Gordy and Bonnie. They had to stop on their way to pick up a lady named Carol that goes to their church, but can no longer drive as she needs a walker to get around with. It was a great time being back at the Duluth congregation. I was thinking that it had to be about 4 years since I had been there! When it was time for me to speak my story was well received, but I was glad to have it over with, as I feel I can write far better than I can speak. I was talking with Burl he said he would be 95 years old in the spring. Burl was saying he had to make it another 6 months as he had some war award he and his son had to get to then he said he really didn't care what happened after that! I was telling Burl that I hoped to see him again sometime but if I didn't I would see him on the other side of things, and we shook hands! After church Gordy had to take me back to his place so I could drive up to Orr before it got dark. I was telling Gordy that I hoped that my stupid pickup would start, it did! So once again I had something to be thankful for. After saying goodbye to Gordy I was off to Orr. It takes about 2 hours to get there form Gordy's place, and the sun was already starting to sink fast! I was going to stop at Mitch and Simi Kennebeck's first as Tom wasn't going to be home just yet. It was pretty dark when I pulled into their drive, but when I entered their home I felt the heat from the wood cook stove, and I felt right at home. There is a comfort in wood heat that is equal to none! Soon after I arrived Tom showed up. After a short chat we all went to his place for supper except for Mitch, he had to work later that evening at the motel so he decided to get some sleep. It was good to be at Tom and Sandy's home again. I always tease Tom saying I have a visual of his place in my mind his mail box is always sticking out of a snow bank! Tom was telling me that he would of liked to burn wood but he wasn't able because of insurance reasons. I was telling Tom that I almost had the same problem once upon a time, but now I don't have any insurance so I can do what I want! That made us both laugh. Ezra Mitch and Simi's son wanted to spend the night at grandma and grandpa's house, but his behavior for the day was not up to Simi's expectations! If he did better tomorrow he might stay, but she would have to wait and see! Sunday morning was a time that I had to prepare for church. Something made me feel that I didn't have enough to read in the time I was allotted. So for the Orr church I read an article I had recently wrote for the Christian odyssey magazine. Tom had already made the bulletins for the day and he had life on the farm for my topic so I felt that fit well, for what I was going to read! After church there was a pot luck dinner I got to meet a few new people that I hadn't met before, as well as talk with those I hadn't seen in a while. After church Tom asked if I would go with him to anoint a lady from his church by the name of Ruth. Ruth had been ill so she wasn't able to be at services that day. When we got to the house we talked with Dale and Ruth for some time, before we anointed Ruth. When the time came for the anointing Tom asked if I would start the prayer, and he would end it. Tom anointed Ruth with oil, and read the scripture from

> ***James 5:14-15****. Is anyone among you sick? Let him call for the elders of the church, and let them pray over him, anointing him with oil in the name of the Lord. And the prayer of faith will save the sick, and the Lord will raise him up.*

I prayed, and Dale prayed Tom said amen. And before we left Ruth appeared to be much better! I was telling Dale and Ruth before we left a story from last fall when I burned my thumb burning some junk. I was walking across the yard afterward it seemed the harder I prayed for it the more it hurt! I had to call Tom for something, and I was telling Tom that I wish he would be here I would have him anoint me. Tom said let's do it over the phone. Tom prayed for my thumb and before our phone conversation had ended my thumb no longer hurt, and there were two blisters on it but no pain! It never did start hurting afterwards. After we left Dale and Ruth's we drove to a friend of ours. He was playing some sort of music so loud that he didn't hear us pounding on his door. We had to climb up on his deck and go around to the front of his house. He has no electricity except a generator I thought I could shut of the generator that would probably get his attention! The house was dark when we pounded on the glass door, and we faintly saw our friend sitting inside he did finally let us in. Our friend didn't seem to be himself he just kept playing this music that was from the 70's and would barely talk to us, and when he did it was just about the music he was playing. I hated to tell him, but I had heard better music! Tom and I both agreed latter that the whole situation was for lack of a better word "weird"! Our friend started to talk he was mussing that God didn't seem to care about him anymore, and that he was thinking of ending it all! I was asking God for the right words, because in most cases when people start talking about ending it all it gives me the creeps! I felt Tom and I could be in a dangerous position who knows what's going to take place? I did ask our friend not in a harsh or condemning way if perhaps he wasn't viewing God in a correct way. That; I told our friend we are probably all guilty of at some time or another in our walk. Our friend answered that it could be! Tom mentioned that we were all friends, and that we could pray for him; he agreed to that! We stood in a circle all three of us with are arms around one another praying hard for our friend. There must have been some sort of spirit bothering our friend, because after the prayer he turned on his lights, and was back to his old self! He even showed us a meteor rock that he had found at one time. That thing must be 40-50 pounds! Before we left our friend was mentioning that he had been very low spiritually lately, and said that God had to of sent us, because if we wouldn't have come over when we did he was afraid of what might have happened! As we were driving back to Tom's place Tom mentioned that our afternoon had been like where Jesus had sent out the 12 apostles, to cast out demons, heal the sick, and preach the good news.

Matthew 10:1: *And when he had called his twelve disciples to Him, He gave them power over unclean spirits, to cast them out, and to heal all kinds of sickness and disease. When we got back to Tom's house; while we were eating supper, Tom was talking about the afternoon we had again. He mentioned that he was glad I had been with him this afternoon. I mention that I was glad to be with him too, what an experience! Tom's grandson Ezra spent the evening that night so we had lots of entertainment. On Monday Tom had to get to work and I had to leave for home another snowstorm was being predicted for the afternoon, so I was hoping to get home before it hit! This time on my way home I didn't stop the engine on my pickup, I just let it run the two times I filled with gas! It started to rain hard when I was half way through the Twin Cities. Why is it that there is always a semi following real close, and the wind shield wiper that is on the driver's side is the one that doesn't work the best? I was glad to get off the interstate at Owatonna, and I was only a few minutes from my dad's house. I stopped to talk with dad for about an hour, because I wanted to get home before the rain started to freeze. When I got home about 4 pm I noticed David Maki had made it back to my place. Again I thought it funny that he spent the weekend here, and I spent mine in Northern Minnesota where he is from! I was thankful to be home as the rain turned to snow shortly after I got home! I spent the most of my time writing this story siting in my recliner in front of my wood stove, and most of the time it was snowing heavy! Wayne Schwartz called me just as I was finishing up writing this story. He told me that he had sold* *his cows, and that we should offer up a prayer of thanksgiving; which we did! I thought what an appropriate way to end this story!*

2 Corinthians 9:11-12: *While you are enriched in everything for all liberty, which causes thanksgiving through us to God. For the administration of this service not only supplies the needs of the saints, but also is abounding through many thanksgivings to God.*

God's peace and a happy Thanksgiving to you all!

Your brother in Christ,

George Denn

Hey By George! February 21 - ? 2011

Hebrews 4:7-14:7: *Again He designates a certain day, saying in David, "Today," after such a long time, as it has been said: "Today, if you will hear His voice, Do not harden your hearts." 8 For if Joshua had given them rest, then He would not afterward have spoken of another day. 9 There remains therefore a rest for the people of God. 10 For he who has entered His rest has himself also ceased from his works as God did from His. 11 Let us therefore be diligent to enter that rest, lest anyone fall according to the same example of disobedience. 12 For the word of God is living and powerful, and sharper than any two-edged sword, piercing even to the division of soul and spirit, and of joints and marrow, and is a discerner of the thoughts and intents of the heart. 13 And there is no creature hidden from His sight, but all things are naked and open to the eyes of Him to whom we must give account. 14 Seeing then that we have a great High Priest who has passed through the heavens, Jesus the Son of God, let us hold fast our confession.*

One of the greatest things about my life on the farm is being able to rest during the winter time. But after several months of resting it sometimes seems more of a forced rest. And before winter is over resting seems to be getting a might old! My thoughts this winter seem to be on the supernatural provisions from God, but I fear in my quest Jesus is allowing me to go through what might be the greatest test of my entire life! I can start my story at this time, but will not be able to finish it as once again I have to await events and how they unfold. My friend Jeff says that whenever God is up to something, something is always going on concerning water. If that is true I will start my story with the day we replaced the faucet on my kitchen sink a few weeks ago. The faucet on my kitchen sink was getting harder by the day to shut off and as usual money had been tight around here so I can't just go and replace things any time that I want I have to wait on Gods provision. I had sold some fire wood during the week, and was going to give the $500 that I made to my neighbor

David as I still owed him some money from last year. I decided that if I didn't get that faucet fixed while I had some funds in my hand I one day would witness water spraying all over my kitchen! So I took 100 dollars of the money and bought the faucet and after another trip to town to get everything I needed to bring my sink into the modern world. While David Maki our winter time guest here was doing some work in the bowels of the sink my eyes spotted the almost gone bottle of soap on top of the sink. There was hardly enough soap left to wash a batch of dishes I verbally mentioned! Well I will just have to wait for some more money to come in until I can buy some I thought. I had just gone into the other room when David hollered "George, come and look what I found under the sink!" Here he was holding an unopened bottle of Dove dishwashing soap we washed the bottle off and it was dated 2-76. I was wondering if it was still good after sitting 35 years under the sink? David thought it should be, as soap doesn't go bad. I was asking David what else he found under there.» Just a couple of old knives, but I left them there!» He said. I know my mom used to buy many bottles of the stuff at a time one must of fell off the ledge and rolled down there long ago. I know I never buy that kind and mom has been away from here for 15 years and dead for about 51/2 years now! I was telling Joe and David that I would call that super natural provision! I mean what are the chances of finding a bottle of soap exactly when you need one?

The 6th annual men's fishing retreat was to be held the weekend of February 11-13 this year. Doug Johannsen had asked me a month earlier if I wanted to go with him. It's a 2 hour drive from my place to Doug's and another 41/2 hours up to Orr where the retreat was. About a week earlier Doug asked me if I wanted to come the evening before, and Visit with him and Betty spend the night and we would leave the next morning. I told Doug I would have to see how my week played out before I could give a definite answer. When Thursday came I was undecided about going up that evening I was thinking I could just as well leave early in the morning, but then I would have the traffic to deal with. I was just going to call Doug I had the phone in my hand, but he called me before I could call him. Great minds think alike Doug said all I did was laugh, because I hope Doug's mind is greater than mine if it isn't we are both in trouble! He was asking if I was going to come up this evening. Sure I said there wasn't anything going on around here today anyway. Doug said I should get there around 6:30 so I said I would leave here about 5:00. It's a good thing Doug called me when he did as it got me out of here as you will see later if I would have stayed here one more hour my whole weekend would have been ruined! After supper and a great evening of visiting with Doug and Betty and a good night's sleep Doug and I left for Orr around 9:30 AM on Friday and we arrived in Orr about 2 PM. The first thing I did was put on my swimsuit and headed for the hot tub! It sure was comfortable sitting in the hot water watching the snow falling slowly outside the glass windows. I was totally relaxed as I wondered what God had in store for us this weekend. I have never been on one of these retreat weekends that I didn't come home more spiritually recharged then when I went. This year especially it wouldn't have taken much! I left my Cell phone in my room and I notice my brother Wayne had been trying to reach me. His message stated that my land lord Bob had been trying to get a hold of me. I called my Brother back and he said Bob seemed a little excited that I hadn't gotten back to him. I found out that He had called about 6 PM the night before. So I had missed Bobs call by about an hour. I asked my brother if he would do me a favor and call Bob back and tell him I was gone for the weekend and that I would call him back as soon as I got home on Monday. Wayne said he would do that. Why is it you can be home and no one ever calls, but as soon as you leave for a weekend everyone is trying to get a hold of you? And why is it that every time the land lord calls I expect the worse? So I had to wonder about the call all weekend, but I wouldn't let it spoil my weekend. If it was bad news I would deal with business when I got back home. We all met at the Dam supper club for our weekend kick off. There was Dale, Tom, Doug, Dan, Gordy, Gary, Ben, Austin, Davy, Alicia, and myself. After a Walleye supper it was back to the hotel for a teaching, prayer and fellowship. Then we retreated to our rooms to sleep. After breakfast our teaching was about the parable of the prodigal son.

> ***Luke 15:11-32:11:*** *Then He said: "A certain man had two sons. 12 And the younger of them said to his father, 'Father,*

give me the portion of goods that falls to me.' So he divided to them his livelihood. 13 And not many days after, the younger son gathered all together, journeyed to a far country, and there wasted his possessions with prodigal living. 14 But when he had spent all, there arose a severe famine in that land, and he began to be in want. 15 Then he went and joined himself to a citizen of that country, and he sent him into his fields to feed swine. 16 And he would gladly have filled his stomach with the pods that the swine ate, and no one gave him anything. 17 But when he came to himself, he said, ‹How many of my father›s hired servants have bread enough and to spare, and I perish with hunger! 18 I will arise and go to my father, and will say to him, «Father, I have sinned against heaven and before you, 19 and I am no longer worthy to be called your son. Make me like one of your hired servants.» ‹ 20 And he arose and came to his father. But when he was still a great way off, his father saw him and had compassion, and ran and fell on his neck and kissed him. 21 And the son said to him, ‹Father, I have sinned against heaven and in your sight, and am no longer worthy to be called your son.› 22 But the father said to his servants, ‹Bring out the best robe and put it on him, and put a ring on his hand and sandals on his feet. 23 And bring the fatted calf here and kill it, and let us eat and be merry; 24 for this my son was dead and is alive again; he was lost and is found.› And they began to be merry. 25 Now his older son was in the field. And as he came and drew near to the house, he heard music and dancing. 26 So he called one of the servants and asked what these things meant. 27 And he said to him, ‹Your brother has come, and because he has received him safe and sound, your father has killed the fatted calf.› 28 But he was angry and would not go in. Therefore his father came out and pleaded with him. 29 So he answered and said to his father, ‹Lo, these many years I have been serving you; I never transgressed your commandment at any time; and yet you never gave me a young goat that I might make merry with my friends. 30 But as soon as this son of yours came, who has devoured your livelihood with harlots; you killed the fatted calf for him.› 31 And he said to him, ‹Son, you are always with me, and all that I have is yours. 32 It was right that we should make merry and be glad, for your brother was dead and is alive again, and was lost and is found.›

The person on the tape did an excellent presentation. Evidently in that time and culture the son that asked for his inheritance was basically telling his father he wished he was dead! Because that is the only time an inheritance was given. In that time and culture being the youngest son his portion of his father's estate would have been 1/3 of all his father's land and livestock and possessions. So his father would have had to sell these things to grant his sons request. Another thing that was interesting was in that time and culture it was the oldest son's responsibility to go find the younger brother and bring him home, something that he failed to do! When the young Jewish boy came to his senses after his Hog raising adventure he went back to his father's place. Let me say from my experience raising hogs it was an experience that won't be forgotten in this life time and probably not the next either, but that's another story! The young man was barely in sight when his dad took off running towards him. Another thing that would not have taken place in that time and culture older men walked they didn't run. His father wouldn't have any of the nonsense about his son wanting to come back as a slave. Instead he dressed him up in fine clothes ordered the fatted calf killed and invited all the neighbors for a party. When his older brother showed up the man was furious, because anything that his father spent on this party came directly out of his 2/3 of his share of his father's estate. Like robes fatted calf's all he saw was this was costing him money! Of course the father came out to reason with the oldest son, but it doesn't appear he had much success at least that's how the story ends. I could see in my own life that I had been rebellious towards God and also legalistic once I believed like the older brother. I also could see that I had been viewing God in a wrong way something that I needed to repent of. I very much want to see God in the way he wants me to know him. Not the way I think I know him! Wow what a revelation! The weekend ended with a church service and potluck meal at Tom's church I think it's called Northland community fellowship. We never did do any fishing this year. In the past we always did that's why it got its name men's fishing retreat weekend, and we did have a young lady this year join us Friday evening, but I was coming home with something far better than fish! I had a better understanding of what Jesus was really like, and I felt at peace something I hadn't felt in sometime. I spent Sunday evening again at Doug and Betty's house than drove home on Monday morning. On entering my house I was met by David he handed me $1000 and said. "Don't tell me you won't take it neither!» I was telling David that he didn't need to do that, but I wasn't going to turn it down neither! Hmmm more supernatural provision! After dinner it was time to call Bob my land lord. I prayed that the Holy Spirit would go before me and be in our conversation. Bobs news was bad! He said that he could no longer rent the farmland to me, because he said his son and daughters were taking over his finances, and they got mad at him because he was renting the farm to me so cheap. They wouldn't even consider renting it to me. How's that for a valentine's day present? Needless to say I was stunned! I had talked to Bob in November and everything was ok at that time I told him I was carrying some guys on some hay and that I would pay as soon as they were able to pay me that was fine with Bob as long as he got it, and I assured him that he would. After I talked with Bob his wife got on the phone and told me that I didn't have the money to rent the farm anymore, and they had a bid from my neighbor Mr. X for $ 225 per acre and they was going to get a contract with him something they never had with me, but that was Bobs Idea years ago. She also told me that she had experiences on this place for far longer than I had as her Father had rented from Bobs dad for a period of 20 years before my folks moved here 49 years ago. I mentioned that I may have to get an attorney to see if it was legal for them to be treating me this way. She kept saying she was through talking with me! I asked her if I could talk to Bob again. I can't remember all the details of our conversation, but Bob told me that I could forget about the $3800 that I owed him for last year's rent, and he said I could live in the house and us the 4.5 acre building site free of charge as long as he was alive. He is also going to rent me the wild hay ground

10 acres for $500 payable whenever I want. He also said I could cut as much wood as I wanted off the pasture ground, because if I couldn't heat this house with wood I wouldn't be able to stay here in the winter propane would cost me about $1000 a month. I burn About 15 pickup loads with side boards of wood here each winter! I told Bob that I wasn't very happy with all of this, but I guess I had to be, I just thought after 49 years things should end on a better note! I do have other ground to raise my pumpkins on, but now it is going to be very inconvenient, and I'm wondering what my other land lords are going to say when they here of all this? After I shook up Joe and David with the news, I spent the rest of the afternoon calling people to pray about my situation. First I called the men that I had just spent the Spirit filled weekend with. First Tom, then Gordy, then Doug. They all seemed very saddened by the news as in many ways the farm had touched their lives as well. Gordy and Tom told me that if I needed a place to live I could stay at their homes for the rest of my life if I needed too! Wow I sure do have some great Christian brothers and sisters! After all the prayers something was telling me that there was more here than what was meeting the eye! If indeed this situation was spiritual it was going to take more than just prayer on my part to move this mountain! My faith at this time seems to me almost nonexistent in spite of what other people feel and that is that I have a lot of faith. Matthew 17:20-21 says.

So Jesus said to them, "Because of your unbelief; for assuredly, I say to you, if you have faith as a mustard seed, you will say to this mountain, ‹Move from here to there,› and it will move; and nothing will be impossible for you. 21 However, this kind does not go out except by prayer and fasting.

If this is so why am I telling God that I haven't the faith for this trial? When Hebrews 12:2 states my faith doesn't come from me, but from him!

Looking unto Jesus, the author and finisher of our faith, who for the joy that was set before Him endured the cross, despising the shame, and has sat down at the right hand of the throne of God.

It seemed that God was also impressing upon me that this was a time for fasting two full days 48 hours. The longest I have ever fasted was 42 hours once about 6 years ago. My natural self just hates to fast, but when God brings you to one its fairly easy! It is very interesting how much clearer his word is during a fast also my prayers seem more in depth! One of the things that God spoke to me near the end of my 48 hour fast was that I should confess all the sins of all the people who ever lived here including my own since the farm was founded much like Nehemiah did at the beginning of his story.

Nehemiah 1:5-7: *And I said: "I pray, Lord God of heaven, O great and awesome God, You who keep Your covenant and mercy with those who love You and observe Your commandments,* **6** *please let Your ear be attentive and Your eyes open, that You may hear the prayer of Your servant which I pray before You now, day and night, for the children of Israel Your servants, and confess the sins of the children of Israel which we have sinned against You. Both my father's house and I have sinned.* **7** *We have acted very corruptly against You, and have not kept the commandments, the statutes, nor the ordinances which You commanded Your servant Moses.*

After that my 48 hours was up I broke my fast. I had made it, funny thing I was never really hungry, and it kept feeling like God was holding me in the palm of his hand! That afternoon I was helping David bring a load of wood into my house before a storm was supposed to hit. My cell phone was ringing it was Blase another one of my land lords. Blase was asking me how it was going and if profits were good last year I could tell that my neighbor Mr. X had been to talk to Blase also, and was trying to rent that farm away from me as well! I never let on to Blase that anything was up or wrong. Something was telling me that I should be completely silent about all this and not say a thing to anyone as to add any fuel to the fire so to speak, and that I should again resume my fasting to keep from these spiritual attacks. By the end of my second 48 hour fast it came to me that I should fast 22 times 48 hours each with 2 meals in between each phase of fasting. What does this all mean? How should I know! I guess God will have to show me, but I guess I will be fasting for the rest of the winter. I think God just answered my prayer to help me to lose weight! I always said be careful what you want you may get it! The thing that stuck out most in this fast was that I should be praying for my neighbor MR. X the guy who is trying to rent my land away from me, and that I should write a letter of apology to my land lord and his wife for some of the words I chose during our phone conversation earlier in the week. Here is a copy of that Letter as I had written it.

Hello Bob and Lorraine

I just thought I should follow up our phone conversation with a note of apology! After I had some time to think about it all I realized that some of my words may have been inappropriate. Although I profess to be a Christian sometimes I fear my actions can be far from that high calling. So if I said anything that may have been out of line I just apologize for that and ask your forgiveness! No one more than I knows how finances can affect our decisions. For myself I have decided to put this all in the Lords hands only he truly knows what›s best for us! Of all the years I have been here, finding the Lord in it all was the best thing about living here! I still am sorry that it went the way it did, but I am very grateful to the Lord that you have allowed me to stay in the house and have the hay ground and to be able to cut fire wood off the pasture! Who knows where this may all lead ? Well just Know that I will be praying for Healing for your Daughter and for your eyes as well Bob.

Best wishes! George Denn
PS. enclosed also is a story that I wrote for my churches magazine that story made its way around the world, and will

be in my next book «Hey by George! III» so God has really used this place to touch a lot of people!

I sent off the letter and then broke the fast. I decided not to fast again until after my friend Terry went home the following morning. Terry generally comes over on Saturday mornings and we go out for breakfast. I haven't told Terry about this nor will I until the proper time! Besides who needs people being suspicious and the last thing I want or need are people's opinions on what I should be doing. I feel Satin can work grandly in that kind of environment, so why even give that being an opportunity to work? The third increment of my fast began at 8 AM Saturday February 19 right after Terry and I went to are usual McDonald's for breakfast. As the third increment of my fast came to a close it came to me that spiritual issues can only be taken care of on a spiritual level hence the reason for fasting! Later that day just as I started my 4 TH increment I read these words from a pastor on the Internet he writes.

"There is a spiritual "world" running parallel to the physical world. We cannot access it through our bodies or souls, but only through our spirit. Your spirit is the connecting link to that other dimension. It has a mind of its own and knowledge that is far beyond that of the mind of the soul. The spirit can share information with the soul, but only as we learn to break down the barriers between them. This is, essentially, how we learn to hear the voice of God, who speaks through our spirit directly to our soulish mind and has even been known at times to speak directly to our physical ears.

My point is that this parallel world is a spiritual world, and it has a direct influence on people and events in our physical world. If you want to change conditions in the physical world (permanently), you must change the conditions in its parallel spiritual dimension, because the physical is the natural outcome of the spiritual."

If I could have put my thoughts to words that's about as close as they could have come! I notice also that I seem to be feeling very excited about my future, because things will never be the same as they were! One thing I have learned about following Jesus these past 16 years is that he doesn't take something away from you without giving you something a whole lot better in the future! And if God would give me back the farm it would never be a land lord tenant relationship! Another thing I am experiencing is peace about all this knowing somehow God is going to handle all of this for me! After I started my 4 TH increment yesterday the food that I ate in the morning was making me queasy. I think I need to allow more time between meals. There is no need to be legalistic about the time frame of this fast. On my way to my dad's this morning I had a strong feeling that I should write to Glenn Taylor, and tell him of my plight. Glenn is a local billionaire and also is the owner of the Minnesota timber wolves. My land lords from my 22 acre Elysian farm know him quite well. I also heard that Bob my land lord here uses the same attorney as Glenn does. This afternoon I had to replace a tire on my pickup at 2: PM on the 22 cd of February what is all this 2 stuff I am seeing in my life all the time? All I will say is this it's how I see God's hand prints on my situation. This is also the second time in my life that I considered writing Glenn Taylor about my situation, and appealing for help. Once about 51/2 years ago also after a time of fasting. That evening I did write a letter to Glenn. I prayed what to say, and even after that I was having trouble though thinking of what to say, it's not every day that I write letters to billionaires asking for help! As morning came by the messages that I was getting I decided that I wouldn't send the letter. The Lord seemed to be impressing upon me that He was well able to handle my situation without me getting involved, and that Glenn Taylor was insignificant as far as this all was concerned! A thought came to me yesterday during the 5 th increment of this fast that was, if we do tinker with things that God is trying to do in our lives, he will not work in ways he would have, if we choose to try and manipulate things! I made a pact with God that I will stay out of things and let him do the things he wants to do in this situation! Today I have peace, joy, and confidence that God will see me through this trial. Monday evening February 28 have just competed 6 th segment of my fast and starting 7 th. At the end of phase 6 I was able to share this story with my friend Jeff he called me during this segment so I felt God wanted him included in all of this. Now that he is the holder of the information that I shared with him he said that now he had to seek God to find out what he wanted him to know at this time. He also shared with me that at this time he had also been in a funk with shallow devotion to God at best. I mentioned that it seemed the way us humans were wired to sometimes get ourselves in these down times with the Lord and God seems to have to rescue us from these times! We started our session with prayer and ended it with communion and prayer! What a way to end a time of fasting with lamb, Lamb of God that is! I Am planning on going up to Gordy's tomorrow probably sometime in the afternoon. There is not much going on around here anyway. As I was finishing up on this segment of this story David Maki said "you know George since you have been doing all this fasting it seems to be having an effect on me also it seems God's word is clearer to me and I am coming to see things like I never have before! Maybe it is effecting him is because he is a member of my household at this time? I concluded the 7th segment of my fast while here at Gordy and Bonnie's place. There are two things that I felt God was revealing to me while I was here so far. The first thing that got my attention was a report that Gordy and Bonnie subscribe to from a William Cambell Douglass II M.D. In his report he was telling how products made from soy that is soybeans are actually bad for you! He claims that digestive disorders, endocrine disruption (whatever that is!) Thyroid problems and nutritional deficiencies all stem from eating too much of this toxic food! I find that all very interesting as it was the first time that I heard that soybean products were actually bad for you. I guess people will have to make up their own mind about such things. I have a theory that it is probably wise to invest in good health by what you eat then trying to correct a life time of eating un healthy by spending what you have made in the medical industry once your health is gone. I heard a sermon 15 years ago now from Adrian Rodgers that the word pharmaceuticals has its roots derived from the word sorcery which by the way was not looked on too good by God! I thought while I heard the sermon boy could that be huge! I feel I must be near a break through raising these pumpkins naturally, and the health benefits that the seeds and oil from those seeds would be to people. That Satin is throwing out all the stops at me like this trial I am in right now that trying to get me to quit raising pumpkins. Given what farmers mostly raise today corn and soybeans and how they are raising them, and all the people and the medical problems they face today; well you will have to decide for yourselves what is good for you and what isn't. All I know is Satin can't win nor will he!

> ***Proverbs 21:30:*** *There is no wisdom or understanding Or counsel against the Lord.*
>
> ***Colossians 2:8-10:*** **8** *Beware lest anyone cheat you through philosophy and empty deceit, according to the tradition of men, according to the basic principles of the world, and not according to Christ.* **9** *For in Him dwells all the fullness of the Godhead bodily;* **10** *and you are complete in Him, who is the head of all principality and power.*

I know what I have done here over the past 16 years I have followed what Got has revealed to me. Also my identity is In Jesus Christ and I have farmed for his glory not just to make money so if anyone wants to go up against that they can just go right ahead, as far as I am concerned! I am more confident than ever that Jesus will take care of me where ever I am! The second thing that I see God speaking to me was through a book Gordy gave me to read titled Have A Little Faith by Mitch Alborn. After I read the book it seemed to me that he meant the book to say have a little faith. Like a host would say have a little wine or have some bread, or have some cake. Not would you just have some faith in this trial! It was such a good book I feel God just served me up a little more faith by reading it. I have decided not to start my 8th part of this fast until I start to think about going home as its too hard to visit and fast at the same time. On Saturday morning I went to church with Gordy and Bonnie. Gordy pastors the Duluth congregation of Grace Communion International. I was there last November, but I could definitely feel the presents of the Holy Spirit moving among the people in a more noticeable way than I ever had before! As the pastor Gordy had been going through the parable of the prodigal son from Luke 15:11-32 for the last few weeks. Immediately after services I was talking to a lady by the name of Joann, before our conversation had ended another lady from the congregation said excuse me George we as a church want to help you in your endeavors, and she handed me a check for $500! That makes $1900 and a bottle of dish soap in the last month that has come to me supernaturally from God that I neither worked for nor expected. Another lady Pam was telling me some ideas that she had about making money farming on a small scale. As well-meaning as everyone is I just feel God gave me the pumpkins to raise for a reason, and I need to stick with that I still have 158 acres to grow them on unless God allows that to be taken away also? Right before we left church there was a book laying on the table Gordy asked me if I had one of those. I told him no, he was looking inside of it to see if there was a name inside, there wasn't so he said now you do! The book was called The Prodigal God- Recovering The Heart Of The Christian Faith. By Timothy Keller. After that we went to lunch it was 3:30 in the afternoon when we started back to Gordy's house. I was starting to think I would leave for home sometime in the morning, so I officially started the 8th segment of my fast. That evening after our game of dominos I started to read the book that was given to me that afternoon. I read about half of it that evening and finished the rest the next morning. As I was reading Sunday morning I asked Gordy if this was the same guy that had the teaching up in Orr a couple weeks back. He said it was. I was telling Gordy that what I was reading seemed very familiar! I was reading along on page 103 and what he was saying about Revelation 22:2 just stuck me like God was saying that is what this fast is all about!

> ***Revelation 22:2:*** *In the middle of its street, and on either side of the river, was the tree of life, which bore twelve fruits, each tree yielding its fruit every month. The leaves of the tree were for the healing of the nations.*

WOW! What a revelation. that what It was all about some sort of healing God was doing in my life as well as those around me! After we had communion Sunday morning I left for home it always seems like it's hard to leave, but I wanted to spend some time with dad and thought I would stay overnight at his place. after I left there's this gas station right before you get on the freeway and I thought I'd stop, and put some gas in so I pulled up to pump 2 and filled it up. But I only got about 2 miles down the road when my vehicle started to spit and sputter. I pulled to the side of the road and called Gordy to come to my rescue! I had visions in my head that I would have to stay longer as it was Sunday and no one would look at fixing the thing until Monday. I also thought of what Bonnie said earlier. «What if you get home and find out God wanted you to stay longer?» Just before Gordy got there I tried to start it again to my surprise it ran fine! Gordy pulled up and I said it was running now. Gordy thought I may have gotten a hold of some bad gas! So we went back to the station and bought a couple cans of heet. I thought it could be a clogged fuel filter, but we put the heet in the gas tank and Gordy said he would follow me as far as Cloquet. The truck was running fine and I didn't have a problem all the way home. After spending the evening at dad's house I thought I would stop at Jeff's church to visit with him. As I drove by the ethanol plant near Janesville I saw my next door neighbor driving his semi coming out of the plant he had just dumped a load of corn there. As I drove by I thought that was like trying to fill a bottomless pit, and is just fueling the greed of most farmers. Jeff and I met at the intersection right in front of the church. Before we started to talk Jeff went to make some coffee and get another space heater for his office. I was looking at the newspaper on his desk front page headlines said, federal support for ethanol may become budget cut target. The article went on to say that the federal support was renewed at the last minute in a tax package last December. It has come under much scrutiny since, and it is making being a livestock producer almost prohibitive! It just seems like God is speaking all through this article and is very significant of what I am going through here on the farm, and what I am coming to with all of this fasting. I shared with Jeff my experiences of the past week afterwards we prayed and had communion. I gave Jeff the book The Prodigal God Gordy had given me. I asked Jeff if he could use a load of wood as I wanted to pass on some of the $500 blessing from the church. he said yes and I hauled it right as I was finishing the 8th segment of my fast. I was also able to get my chainsaw back from Jeff as he no longer needed it because of the gift. Now I can start cutting wood again for next year I guess! I just started the 9th segment of my fast today. I am starting to look forward to what God has to reveal to me during this segment of my fast? Jeff called me this morning thanking me for the load of wood. It was quite a pile he said! He also thanked me for the book and said he almost had it read. He said he could see the Holy Spirit all over it. Later this morning I was hauling a load of wood to my accountant and I realized that the end of my 9th fast would end at 9pm on the 9th of March, and I no way planned it that way! Again I could see Gods' hand prints on this fast. As I finish with my 9th fast I met with Tim and Bob today down at kids against hunger for prayer we also had communion one of the things I noticed on the way into town was my neighbor Mr. X pulling out of another of my land lords drive ways, at first my heart sank but not for long we prayed about that I have to let the Lord handle things, and for me it's time to eat!

As I start the 10th part of my fast once again I find myself wondering what God will revel to me, and what he has in store for me in the next couple of days. I feel I understand now what James meant in his epistle when he told us to count it all joy when we are going through a trial!

> ***James 1:2-4 2:*** *My brethren, count it all joy when you fall into various trials, 3 knowing that the testing of your faith produces patience. 4 But let patience have its perfect work, that you may be perfect and complete, lacking nothing.*

Because up until now except for a few fleeting moments, I have just felt joyful most of the time! I keep wondering if it is God holding me up or am I just in denial? I guess I'll just have to wait and see. Yesterday I was visiting at my Dad's, and a couple of his friends showed up who go to A.A. Meetings with him. I have been praying for Dad about a situation in his life during this fast. His friend was telling Dad that he could help him get signed up for veterans benefits so he wouldn't have to buy insurance anymore. The money he would save would allow him to stay where he is a little longer. without going into any more details this would be an incredible answer to prayer! On my way home from Dad's place I was praying for God to send buyers for the fire wood that I haven't sold yet about 2 seconds after I said amen a guy buy the name of Randy called for a load! It turned out that he was an older brother to someone I went to school and graduated with. Before I hauled the load of wood I noticed Joe had a pouch of almond pipe tobacco laying on the dash of his car. God seemed to be impressing on me that I was allowing tobacco usage on my place when he had convicted me some 16 years ago that it wasn't any good, and I find it hard to believe that he would want me to quit using it , but didn't mind that others did! I have since climbed out of that life style, but I was promoting it if I was allowing it so I knew I had to talk to both Joe and David. my chance came after I returned home from delivering the load of wood. when I walked into the house Joe and David were eating lunch. I sat down in my chair and asked them if I could talk to both of them about an issue without them getting mad? They said, "yes." I said 16 years ago when I gave my life to the Lord one of the first things he convicted me about was my tobacco habit, but I see tobacco on the dash of Joe's car and David going out to his truck all the time to smoke cigarettes. I told them that I didn't want to see the stuff anymore or see any one that lives' on the farm here use it. What they do when they are gone from here is none of my business, but I don't want to promote it here, and if I see it I am allowing it, and if I am allowing it I am promoting it. And I asked them please to stop. Joe said Ok and shortly after went out and put his tobacco and pipe out of site. David said that he knew he was pushing the envelope buy smoking in his truck . Since then he takes a drive when he wants to smoke. Tobacco situation dealt with! Today Wayne stopped by to pay me for the rest of the hay he had bought last year. I also had a meeting with the 2 Hy Vee grocery store managers about buying my pumpkins and squash next fall. They contacted me, and it all sounds very promising provided I have some ground to plant them on come spring! I picked up my mail this afternoon after the appointment with the managers of Hy-Vee. I almost walked out of the post office when the lady at the counter mentioned that she had a package for me. I thought it may be some more pumpkin seeds that I had ordered. Turns out it was from Sally Limbeck from Bend Oregon. I Met Sally about Three years ago to the day that she wrote the letter 2-22-11! Sally works at Whispering Winds retirement home in Bend Oregon, and my sister Marie works there as well. Sally was mentioning in her letter that Marie had given her a copy of my second book Hey By George II for a gift for Christmas. Along with her note she sent me a can of assorted candy bars but at the bottom was a small brown pill bottle and inside was 6 $2 bills from 1976! Needless to say if you are familiar with the story that I wrote about $2 bills from 1976 on pages 51-53 in my second book you would know why these encouraged me greatly! You can also read about my trip to Oregon in February of 2008 to visit my sister, and how I met Sally on pages 80-87. Well I still have 14 hours to go before I eat anything so will have to see if there is any more to write about.

Almost through the 11th segment of the fast after I end this segment I am half way done with what God asked me to do! I stopped again at Jeff's church we had communion and prayer again. One of the topics we discussed was the earth quake in Japan it was now upgraded as a 9 on the Richter scale! One of the biggest since 1900. After that I hauled a load of wood, and ordered some barley seed for the organic farm 76 acres will be in barley this year unless of course God has other plans? A few hours ago the Le Sueur Co. Sheriff Stopped and gave me some papers that I have to appear in court on the 21 st of April apparently I was supposed to fill something out and send it in and I didn't For some collection agency back in November. they are taking me to court the way it looks to find my reason why I didn't send it in! the reason is that I burned it! so It looks like God will have to work some more on this case! The day I go to court is the day after my 49th birthday I hope this isn't like the valentines present I received! Well anyway like I always say thing always tend to get interesting when the cops show up! It looks like I can eat in about 2 hours I guess that's all I have to write about this time. On a parting thought when I started this fasting I weighed 235 pounds I just weighed myself and am now 218 pounds. So in the words of John the Baptist I end the days writing.

John 3:30: He must increase, I must decrease.

March 17 St. Patrick's Day! As I near the 12th part of this fast it seems like I have a war going on in my head. On the first half of this portion of fasting I found myself doubting myself that I had enough faith to go through this trial, and to finish the fast that I am on! I had to ask Jesus to help me in my unbelief just like the guy did in Mark 9:23-24:

> **23** *Jesus said to him, "If you can believe, all things are possible to him who believes."* **24** *Immediately the father of the child cried out and said with tears, "Lord, I believe; help my unbelief!"*

And do you know that seemed to help! Shortly afterward I sold a load of firewood, and from some things I had been reading, and an e-mail from a friend, and a call from another friend I could hear Jesus speaking encouragement into me! I find I am encouraged to move forward. I can't wait to eat again, but I still have 3 hours to wait! Phase 13 of the fast! I am still going 2 folks told me that they never heard anyone do what I am doing. One of those guys was Joe my hired man. I reminded Joe of a statement that hangs on my bedroom wall says. «If you want to see what you never have seen before, then you must do something you have never done before!» 2 things interesting happened today. the first thing shot out at me while I was reading Joshua 9:3-5:

> **3** *But when the inhabitants of Gibeon heard what Joshua had done to Jericho and Ai,* **4** *they worked craftily, and went and pretended to be ambassadors. And they took old sacks on their donkeys, old wineskins torn and mended,* **5** *old and patched sandals on their feet, and old garments on themselves; and all the bread of their provision was dry and moldy.*

It came to me that how often Satin works very craftily trying to enter our lives making himself even appearing as a friend at times!

> ***2 Corinthians 11:14-15:*** *And no wonder! For Satan himself transforms himself into an angel of light.* **15** *Therefore it is no great thing if his ministers also transform themselves into ministers of righteousness, whose end will be according to their works.*

Shortly after that David Maki and I were talking in my kitchen. David was asking me if it was ok if he burned something in my wood stove in the kitchen. (My wood stove has a name Terry and I jokingly named it hell fire and the one in my basement is called damnation!) David told me that this was the day that he was going to quit smoking cigarettes and he was wondering if I would let him burn them in hellfire? Of course I would, I had been praying for him and this habit since last September! David through those things into the fire and Joe and I laid hands on David and prayed for him for Gods strength at this time. I also read David a promise from God to him from 1 Corinthians 10:12-14:

> *Therefore let him who thinks he stands take heed lest he fall.* **13** *No temptation has overtaken you except such as is common to man; but God is faithful, who will not allow you to be tempted beyond what you are able, but with the temptation will also make the way of escape, that you may be able to bear it.*

I was telling David that scripture would be pretty important to him in the near future. I feel tobacco is a habit strait from hell, and it is an area in one's life that one gives over to Satin, and Satin doesn't give up on you without a fight, because Satin doesn't like being evicted!

March 19

Just today during the 13th part of this fast my other landlord Blase called me to see if I had some hay. He was the guy who called me a month ago to see if everything was going ok with me! I told Blase that I still had some hay and he mentioned that he would be right over. I was starting to feel a little nervous as I thought Blase would be full of questions for me! A minute after Blase called anther guy called for a load of firewood his name was Mike. Blase pulled in with his brother in law just as David and I were loading the load of wood. I was praying that I wouldn't act peculiar or say the wrong words to Blase finally I told Blase that I had to go and deliver that load of wood. I told Blase to just help himself with the hay the price was $3 a bale and if it was alright with him we would just take it of the rent. Blase told me that would work the best for him so it looks like I have that farm rented for the season, and Mr. X's attempts to get that farm away from me were futile! So God has given me a direction it looks like he still wants me to farm on some sort of scale so that encourages me! Also after 16 years selling firewood in the Mankato area I never delivered wood to 6th street until just a couple of weeks ago and today and they were only 18 house numbers apart. Mike new the other guy but didn't have much good to say about him although both parties were an answer to prayer, perhaps one of our humanly issues is that we don't treat our fellow man in a proper way that God would have us do? Here is something interesting that I just read that also encouraged me. «We may well be entering an unprecedented era of natural and man-made disasters. These, of course, were predicted long ago in Matt. 24 and in other places, so we should not be surprised.

Neither should we be afraid. The Scriptures repeatedly admonish us to «fear not.» In the face of danger, we are to have faith, not fear. Fear compromises our immune system. Faith strengthens it. We can measure our faith by the level of JOY that we have in the face of fearful circumstances.» Joe is cooking one of the turkeys he raised this year as his birthday is tomorrow the 20th it sure smells good! Only problem is I still have 3 hours to wait before I can eat!

March 20 evening 14 part of fast. After praying for what God wanted me to see and know from the scriptures this evening he gave me a revelation to a dream that I had several years ago! I had not thought of it for some time, but as I was starting to read 1 Samuel I got as far as when God was calling Samuel in chapter 3, when I had to go back and re-read part of the previous chapter. As I was reading God just put that dream into my mind and reveled to me what it meant! One of the interesting things was where I started reading was in 1 Samuel 2:22, and I didn't notice that until after the fact! Here is the revelation as it was given to me. In my dream I was walking in a woods every now and then I could get a glimpse of what looked like a river off to my right. the woods in this dream represents my former life before I began to follow Jesus. I walked straight out of the woods I could see not only one river but 2 rivers running parallel to one another with a small strip of land in between somewhere between 20-40 feet, but I couldn't get to them as there was a fence with a lot of dead weeds alongside it tangled in between the wires. This was the time in my life that God was calling me, but my own pursuits kept me from getting where God wanted me to be. The time of year was early spring just as it is now with snow piles here and there yet that hadn't melted and the sky was over cast even slightly foggy typical early spring weather. When I came to the end of the fence I immediately turned right 90 degrees and walked straight across that first river the ice was very solid yet, and I felt very safe walking on it. After I was across I spent the night on the other side of the river. When I woke up in the morning I was surprised to see that the ice had broken up as it was well froze yet the day before. I was standing near the edge of the river looking very close at it. There were chunks of ice floating everywhere very close, but the river didn't seem like it had any currant to it! In between the ice chunks were many fish, but they were all the same size and all the same kind They appeared very helpless and some were looking at me, but it seemed like I was helpless to help them also! The point where I tuned 90 degrees was the time I finally gave my life to the Lord and decided to follow him The first river represents the church for a time it was a safe place to be, and the solid ice was a safe place to walk! Waking up represented an awakening in myself just as I was surprised to see the ice brake up in my dream, so was I surprised to see my church brake up in real life and the chunks of ice not only represent my church, but others as well. And the fish represent the people in those churches they were unable to help me and I was unable to help them. One must be part of a church brake up to understand the sadness of it all! I also had to take a new direction in the way I understand what church means for myself. I now understand it to mean now I am the church (the ekklesia " called out") called out to go out into my world and be the church! I still am connected to the greater church and often participate in what I call intense church, but I have no weekly meeting place. After this I walk across a narrow strip of land that slightly curves up and slightly curves down to the edge of the next frozen river. I find I identify with this strip of ground as a farmer the land is my lively hood and since my church has broken up there was a gradual financial climb so to speak and a gradual financial decent so to speak, but alas my identity can no longer be in the land or a farmer it has to be in Jesus Christ and he represents the sun that was getting hotter and hotter as I was crossing the next river! So hot that the very ice I was walking on was melting faster than I could walk so I had to start running, but the melting was taking place faster than I could run so I had to leap for shore that was very far away. Farther than I ever had to jump ever! God revealed that that's where my life is now in that leap of faith! Jesus is allowing me to be stretched farther than ever before. Will I make the leap? Not without total reliance on him. I woke from the dream not knowing if I made it to shore. I sit here today not knowing how this story ends! Finally an interpretation to that dream that fits, because it was God who gave me the dream about 6 years ago now, and it was God who just gave me the interpretation!

Monday March 21 went and told Jeff about the goings on these past few days we had prayer and communion. As I write this I have 12 hours left of fast 14 will end it 7:05 am in the morning.

I am now on the 15th part of this fast. God continually show me something in each fast that I can tell it is him! Today my friend Bob had called me around 10 o'clock am. Bob is a friend that I pray with occasionally he knows my situation here. He had a booklet that someone had given him, that he thought that I might get something out of as an encouragement at this time. He said he would leave it with a friend in town so I could pick it up at my leisure. For no reason Bob suddenly changed his mind and decided to bring it out to me immediately! After Bob showed up he was telling me of all his health issues! I was telling Bob that he ought to try some pumpkin seed oil. I was telling Bob how good I felt after taking it for about 5 weeks. I was telling Bob that since I have been taking a teaspoon a day, I don't have pain in my joints at night so much anymore. Also my seasonal depression seems to have lifted some time ago which is odd given our weather hasn't settled yet! It also could be from all the fasting I have been doing, as I have lost 20-25 pounds from where I had been weighing earlier. That in itself makes me feel much better! I was telling Bob about my plans to crush my own pumpkin seed oil this fall out of the left over pumpkins. Everyone I talk to seems to be excited about my pumpkin seed oil venture! After a time of prayer with Bob as he was leaving a UPS driver appeared with some pumpkin seeds he was delivering that I had ordered earlier this winter! The booklet Bob dropped off was named Someone Cares. I read it and passed it on to David hopefully he will find some encouragement within its pages at this time he is quitting smoking? Today I am also getting ready to leave for Illinois to take Dad down to visit his cousin while I continue on to take in some summer camp training down by Lacon IL. I will stay at Dad's tonight, and God willing we will leave first thing tomorrow morning, and be at my cousin Nancy's around 4 PM tomorrow evening!

Our trip down to Nancy's started about 7:45 am on Thursday morning. We drove the 480 miles in exactly 8 hours. It seemed about to be our fastest time ever we made it down there. Dad and I talked the whole way there, we haven't had much time this past winter to do that so like they say nowadays we had some quality time together! One of the funny things that happened on our way down there was when I ended my 15th part of my fast I was only 8 minutes to Princeton IL. a town that I always stop to eat at McDonalds on my way through. I started thinking how good an Angus burger and fries and vanilla shake would taste! when I drove into town I couldn't find the place which was odd as it was just off the freeway. Thinking I had just somehow driven by it I turned around only to find a pile of dirt and 2 bulldozers where McDonalds used to be! So we had to drive over to Culver's instead. I had a triple butter burger and large vanilla shake to end my fast. Mmm mmm! After we got to Nancy's house we drove back to Coyote Canyon in Bourbonnais. One of my favorite places to eat when I am down there. After we got back to Nancy's it was getting late so I was thinking of taking a walk so I thought I would do one the next day. I woke at 7 am it was sure nice to see green grass and trees starting to bud, because when we left my Dad's house an inch of snow had just fallen so the green was a site for sore eyes! Nancy had plans to take us over to Indiana to a town called St. John's. There is a place there called The Shrine of

Christ's Passion. It's a gift shop and outside they have a trail that is supposed to be pretty identical to the original path that Jesus took to the crucifixion, and all along the way they have life size bronze statues from where the journey starts at the last supper. From there you enter the open gates to the garden of Gethsemane where you see the disciples sleeping while Jesus is off yonder praying. On to Pilate where he was washing his hands and siting on his stone chair. I sat down beside the old boy and could understand why he was such a hard you know what! If we sat in a chair like that all day every day we would probably be one to! So I thought I'd have some fun with old Pilate to help him liven up a bit, so I put my hat on him and had Nancy take a picture of us. As our walk went on through the various stations I could feel the presents of the Holy Spirit. It was interesting to take this walk with my Dad and our relatives Donna and Nancy. As I stood under the cross that held the bronze image of Jesus Christ with the 2 thieves that were crucified on either side, again Nancy snapped my picture as I looked upward and meditated on the scene. We walked on to the tomb that was of course empty except for this bronze angel that was kneeling in there. As I was walking out Nancy said go back in so I can get a picture of you. Latter when we looked at those pictures there were three circles all intersecting in a vertical line all the same size! Nancy thought that they may have been angel's near me, but I felt it was a sign that our triune God was right there near me! Then on to the most important fact of all where Jesus was ascending back to his heavenly Father to sit at the right side of God Almighty where he still is to this day interceding for us humans day and night! Wow! I wanted to take a walk today , but who would have known it would have been the ultimate one? It sure helped me to re-evaluate what I believe in, and how far I have come! After that we had a shake at McDonald's, and Donna and I had our picture taken with the purple Grimace that was waving at everyone in front of the store. We drove on to a town named Lowell it reminded me of my Uncle , because he has the same name. We drove back to Nancy's house and grilled some hamburgers I had just about an hour before I would leave for camp training and I was starting to get excited about the weekend that lie ahead! I was wondering what God had in store for us that were going, and excited about seeing some good friends that I haven't seen in a while! My drive to camp takes me across Illinois hwy 17 one of my favorite roads to drive on! I feel it is rural America at its finest. As I drive I notice field work has started here and there across the state for another year it's a little early, but I imagine in a couple of weeks it will be going in full swing! For some reason my drive felt different today for reasons I can't explain, not a bad difference, but just different! Perhaps it is because the fields are so bare yet? Words can never express the greatness you feel to be reunited with these wonderful people that I work with year after year! We all come together to be part of something God is doing with us and the youth year after year! after a light meal some praise and worship and some instructions we were free the rest of the evening to talk' play cards or play other games which for me was until 11:30 PM. Saturday started with breakfast at 8:30 AM. With praise and worship and a speaker until noon. After lunch there were speakers and a few activities until the evening meal. it was interesting to listen to the speakers discuss a lot of the things the Lord had been talking to me about in the last couple of months! After the evening meal and towards the end of our card playing I was feeling a little on the anxious side no longer did I feel like the Lord was holding me in the palm of his hand. This was a feeling that I had rarely felt these last couple of months during this period of fasting, and prayer and my confidence in him was fading! I asked David

Holmes if he and several of the others would pray for me, and the situation I was in. David Holmes, Mike Frederick, Kevin Ford, Todd Woods, and Jordan Rex all laid hands upon me and prayed each in their own turn. At first I just felt their hands on my back. Slowly I could feel my back warming until it felt almost as hot as fire I could tell that the Holy Spirit was present during their prayers! When the praying ended I made the comment « wow what an intense experience!» Little did anyone know but it effected Jordan who is 18 and will be joining us on staff this year. He latter opened up to David Holmes about some issues he was having in life, because of witnessing what went on with me. I always call that sort of thing the deeper things that go on at camp! After David told me that I was thinking that if what I was going through in life went to help someone else than it was well worth going through! I was telling Kevin and David latter my thoughts. I also commented that one could search the whole world over, and not find a greater quality of friends then those people were that were in that room this evening! as the weekend came to a close I left the camp at 8:30 this morning Sunday March 27 and started my 16th fast as well! as I drove the many miles back home the time just flew, and my thoughts went back over the experiences of the past few days. I also began to wonder what God had in store for me when I went home and my near future/ I guess that one as it is right now will remain to be seen!

Well here I am just 2 hours and 14 minutes from finishing my 17th part of the fast! I can safely say I am hungrier than I have ever been in my entire life! I started cutting firewood a few days ago, and I am finding that working without eating is extremely hard. I had to quit today about 1:30 as I started to get weak, but spiritually it is well worth it. Between the end of my last fast and the beginning of this fast I was invited to the home of a good friend of mine for supper. Just before I left my friend Jeff called me and said you know George I think you are on the right path with this pumpkin seed oil idea of yours, which to me the timing was all too perfect to be coincidental so I was greatly encouraged! After a supper that was out of this world! One of the biggest things I can see is how grateful I am to be able to eat, and how well it all tastes since I've been on these fasts. All I can say it brings thanking God for your food to a whole new level! Later that evening I had the opportunity to pray for the Lady of the house she had been sexually abused by her father, and is having an extremely hard time forgiving him for what he did long ago. She knows that the forgiveness that she needs will have to come from Jesus himself as it is not humanly possible to have that sort of forgiveness come from within one's self! What is so sad not only did this person abuse his daughter, but this family was also a part of the church. It sure was a touching experience to be able to pray for this Lady. While I was sawing wood the next day I was wondering how many people I have wronged in my lifetime that I am not even aware of, or don't even remember? How desperately we need Jesus to work out and act in our lives! I was just sitting by my wood stove this afternoon sort of counting the minutes until I can eat again when Ernie called and said he wanted to stop out I was glad for the company. After a while my friend Jeff stopped out as well. I had Jeff bring me out a copy of Rob Bells latest book LOVE WINS as I here it is causing quite a stir in the Christian community. I don't know why I like people that stirs things up perhaps I am one of them! One of the interesting parts of our conversation Ernie said he couldn't figure out why God would sent an evil spirit into King Saul?

> ***1 Samuel 16:14-15:*** *But the Spirit of the Lord departed from Saul, and a distressing spirit from the Lord troubled him.* **15** *And Saul's servants said to him, "Surely, a distressing spirit from God is troubling you*

Jeff's answer was an interesting one. He said there was 2 things going on there Saul was jealous of David and instead of admitting the sin he justified it, and God was taking responsibility for Saul's sin by sending the distressing spirit into Saul! I thought that was a very interesting answer. After Ernie left Jeff and I continued our talk and had some time of prayer for all these goings on, but most of all that I would finish what God has brought me too and to finish strong. When I did eat I was so very thankful. For me being thankful for the food that I receive has come to a way higher level!

6:46 a.m. Monday April 4 1 hour and 44 minutes left to go on my 18th fast! Last week when I was at camp training I was thinking that these last 7 times of fasting would get a little crazy but so far it has been pretty tame, but who's complaining? God sure has been speaking to me in small ways which is just as important to listen to as in big ways. I figure if I can't hear him in small ways, how will I hear him in the big ways? Yesterday I traded my neighbor David 2 Red Oak logs for a quarter of beef. David also has chicken, and they are laying more eggs then they can use or sell so every week he's been giving me 4 dozen eggs. Darcy, David's wife works at a bread store and she brings home the bread that is past the date that the store can sell. So every week I get all the free bread that I can use. David also Lets me get the Diesel fuel that I need for farming and he lets me pay for it during the fall pumpkin season when I have the money! So I am very thankful for my neighbor David, and I pray God blesses him as much as he has been a blessing to me! I stopped by the organic farm that I run I saw Doris my land lady and we talked about everything from watermelon pickles to pumpkin seed oil! I am so thankful for Doris. God gave me that farm to run 11 years ago now. I does amazes me how one person can be so for you yet another can be so against you!

I spent yesterday moving some machinery of the fields out back and into the yard I would have to do this regardless who runs the land. I also managed to get one of the chimney caps put on the house yesterday, as I write this it is raining so I am thankful to have it on, because if it rains to much that chimney will leak! Last evening just after I went to bed Ernie called me with a biblical question from 1 Chronicles 4:21.

> ***1 Chronicles 4:21:*** *The sons of Shelah the son of Judah were Er the father of Lecah, Laadah the father of Mareshah, and the families of the house of the linen workers of the house of Ashbea;*

Ernie was telling me that he had just read where God had killed ER, because of some sin and if that was true what's ER doing alive again? I told Ernie to just hold on until I could get up and get my bible and read what he was reading maybe I could answer his question. The answer is simple Shelah, Judah's 3 son did marry Tamar after the scandal of her and her father in law which produced Perez and Zerah! Perez being in the lineage of Jesus Christ. After Shelah and Tamar married they had a son that They called Er to carry on Tamar's first husbands name and linage. Hm interesting I never caught that before I always thought Judah somehow took Tamar in as well as the two boys that the scandal produced. It also coincides with my friend Jeff's theory that God takes responsibility for our sins! Thank God for Ernie getting me out of bed last night and asking that question! Well this fast seems to be about being thankful, as well as being aware of the little things, because a whole bunch of little things generally add up to big things!

> *Luke 16:10: He who is faithful in what is least is faithful also in much; and he who is unjust in what is least is unjust also in much.*

After I had a good breakfast I drove over to talk with my friend Jeff at his church. As usual we talked about the week's events and how we perceive them, and what God had been speaking to us in general. As we talked we had 2 unexpected visitors. They left just before we had communion both Jeff and I thought we should have invited them for that it was interesting how we were thinking the same thing. Jeff was asking me if I had a theme perhaps I had for communion I thought a bit and I responded thankfulness. Thankfulness for all the blessings God had given me to be able to pray for those individuals that put me into this trial. Thankfulness for those that are for me, and are praying for my situation. And thankfulness for all God has shown me as well as sustaining me through these times of fasting! Jeff said it was interesting because he had his bible open to psalm 105 for the past 20 minutes he read verses 1-5.

> ***Psalm 105:1-5:*** **1** *Oh, give thanks to the Lord! Call upon His name; Make known His deeds among the peoples!* **2** *Sing to Him, sing psalms to Him; Talk of all His wondrous works!* **3** *Glory in His holy name; Let the hearts of those rejoice who seek the Lord!* **4** *Seek the Lord and His strength; Seek His face evermore!* **5** *Remember His marvelous works which He has done, His wonders, and the judgments of His mouth*

I started fast no.19 at 6:00 a.m. it's now 9:50 a.m. Tuesday April 5 only 4 more fasts to go! I wonder what God has in store for me this time? Wayne Schwartz a good friend and fellow farmer stopped by last week, and mentioned that he had been invited to a meeting for fellow Christian farmers. Wayne was asking me if I wanted to go. I told Wayne that I would ask the Lord if I should as maybe I would be too controversial of a figure there. The question they were asking at the meeting was as Christian farmers how do we treat one another and how do we operate our businesses? Now to me that question is a no brainier so one of the reasons I didn't want to go was for one I have had about enough meetings of this nature to last a lifetime and 2 I didn't want to go to this and state my feelings and tick every one off! I told Wayne that I would have to think on it a spell. As I was cutting fire wood I was asking God if this meeting was something that he would want me to go to? About Tuesday evening God told me that I should go to this meeting. I thought I will just listen to what is being said, and speak if only I was asked. My only thought was I hope that I don't go to this meeting and find out later that I should have stayed home and got my wheat planted! The morning of the meeting was a day I could eat so I was sure thankful! I weighed in at 204 pounds that morning which makes me feel even better because up to this time I have lost 25-30 pounds! Wayne, Joe and myself went to the meeting Wayne suggested that we pray before we went into the meeting as none of us knew what to expect! It didn't take long listening to the folks around the table that all of us were there because God wanted us to be there! Many of the guys were talking about the very things God had been impressing upon me. We were all asked the question what we thought God wanted us to say. I just briefly mentioned about the fast God had me on, and that if they would just listen to what God was saying to them they needn't worry about a thing as he would get them where they needed to be. One of those farmers was Dan French a dairy farmer down by Kasson MN. I had meet Dan about 6 years ago at the Church of Christ there when I was introduced to the market place ministries. I was able to talk to Dan a couple of times just briefly. I told Dan of the fast that I was on. Dan told me of the 40 day fast that God had put him on one time! He said he to had lost a lot of weight and since that time his legs were always cold! I explained to Dan my fast and the purpose of it. Dan thought that God was going to do something really good in my life. I always take what people say with a grain of salt, but what Dan had to say greatly encouraged me. that evening when we got home from the meeting Joe David and I went out in the woods to split some more wood. David and I went out first as Joe had to do a few things before he came out. I was moving some big chunks up to the wood splitter with the skid loader when I noticed my chainsaw was not where I had left it almost immediately I heard this crunching sound coming from down by the wheels! It didn't take me long to figure out I had just flattened out my chainsaw! It was un-repairable! So being I have no money to buy a new one it looks like I am out of the wood business until God provides for a new one! Shortly after that my axe handle broke since I came home from the meeting it has not been a good day! Friday dawned and I thought I would go to Blase's to see if the ground was dry enough to plant wheat? It was dry enough and I got the small field worked and planted by 1:30 in the afternoon. I had to go pick up Dad and Donna around 6 o'clock in the evening, and they wanted to take me out to eat. Also Terry was coming in the morning to eat breakfast so I thought there would be no use fasting until after breakfast on Saturday. Terry called me on Friday evening wondering what I was up to? I told him nothing as I had run over my saw so I couldn't do anything! he just laughed and said he would float me an interest free loan for a new chainsaw until the pumpkin crop was selling! I figured that was a God thing because I never asked him he just offered! So in the morning after breakfast we went to Smith's Mill implement and bought a new chainsaw! Jim Bowers the sales man and half owner was telling us that was the first sale of the day! when we went up to the counter to get the sales slip Ed Mulcahey the parts man was there Ed has been a parts man there ever since I can remember! As usual we like giving one another a hard time. I was telling Ed that if he didn't stop giving me such a hard time every time I came to get something there I would have to write about him some day! I guess he just pushed me over the edge this time! I was saying that this would probably be the last chainsaw that I would buy everyone just started laughing! Jim said I think I heard that comment before Terry said yea 2 chainsaws ago! Ok we got to get going we have some work to do, we can't just stay here and palaver all day! I was thinking that maybe we could get the couple of acres of wheat planted at David Maush's place done today before it rained? Terry and I stopped at Dave's to burn a brush pile that was sort of in my way. Once we had the brush pile burning David had a few other trees near the pile that he wanted to trim. One fell on me while I was cutting it and knocked me down and knocked my solder out again! I was glad it wasn't any worse. I did get 3/4 of the wheat planted there was still some frost in the ground near the edge of a grove I had to leave until another time. My 20th fast started right after Terry and I had breakfast at 7:30 this morning. For some reason I feel an urgency to get my fasting completed as quickly as I can! It is a feeling I haven't had at all during all this time until now, so I wonder what's up with that. This fast will be done on Monday so I will only have 2 more times after that. David Maki mentioned later that day that he planned to be leaving for the North Country on Wednesday

Morning I sure will miss Dave around here, and let him know that he is more than welcome around here anytime. I was telling David that I have told a lot of people that over the years I just hope they don't all take me up on that offer all at once, but if they did I guess we would have quite a party! I keep feeling this urgency to finish this fasting as soon as possible perhaps God has something coming that we cannot see at any rate I should finish sometime on Saturday morning!

11 hours into my 21st fast, and with an ongoing urgency to finish as soon as possible! For some reason I am extremely tired this morning, perhaps there is a lot going on in the spiritual realm that I am not aware of? Yesterday just before breaking the 20th fast something came to me what I should do. Yesterday which was Sunday I was up to my dad's house and I was looking at a newspaper that Donna had picked up on Friday, they had kept the thing because my dad thought his sister's picture was on the front page. Looking closely at the picture any one that knows her can see that it is not her! After I had established that fact I was paging through the rest of it, I noticed Bob my land lord here his daughter who is named Donna too had died from cancer. I thought I should send them a card with perhaps $20 in it, but this morning God seemed to be impressing on me to also send one of those 1976 $2 bills that came from Sally in Oregon! I had been praying for Donna that she would be healed, I guess being with Jesus is the ultimate healing so it looks like God answered my prayer. I went to get a card on my way to see Jeff on Monday I put the 22 dollars in it and wrote something encouraging for the whole family, and sent it off. Jeff and I didn't have so much to talk about this time we went to see the spot where

I will plant oil seed pumpkins on. Then we came back to the church and had communion then I had left I was supposed to meet Dad and Donna around 11 o'clock, but they were already at my place ,and it was only 10 o'clock. On my way home from the church a guy called about some hay, he had called a couple weeks back. I had him give me his address so I could get back to him with what the hauling charge would be. It was interesting because when I typed his information into map quest the time of travel from my place to his was 2 hours and 22 minutes! Hmm where have I seen those numbers before? It is interesting as this is the only guy who has responded to my hay ad all winter except for the guy who calls me every month to renew the ad! But the best thing is that he just called back and for sure wants a semi load which will be about $2200 which at this time I desperately need! Thank you Jesus! Yesterday I was also able to get the clover seed that I needed. I just called my seed guy Gene Werner and said «Gene I need about 1600 pounds of Arlington red clover seed and if you can carry me until August 1 or so that would be great!» Gene said «sure I can work with you on that.» So it looks like things are shaping up for me this spring. I just wonder what for sure is going to happen at this place, but I expect I will find out shortly! Will finish this fast Wednesday evening at 9:51 PM I will be glad to finish this fasting but will miss the closeness to God I feel during fasting, it is like none other! An added bonus I have lost 30+ pounds.

Wednesday April 13 5:55 PM with less than four hours to go with this portion of my fast I can continually see how God provides for my immediate needs. I was telling Wayne Schwartz that I was going to go to Albert Lee one of these days to pick up my barley seed. I was telling Wayne that I intended to pull a hay rack down and load the seed on it. Wayne was saying that was a long way to pull a hay rack he needed to go down there also so we could go with his pickup and 5th wheel trailer. Wayne was telling me he would be at my place at 1:00 this afternoon, but he showed up at 12:40 this is so not Wayne that it seems miraculous! Wayne is always an hour late always! Wayne was also in a hurry to get back home by «no later than 5:00 PM» his home that is, 20 miles from here. It takes about an hour to get to Albert Lee from here Wayne seemed quite talkative while we were there. Wayne felt that we had plenty of time to get back, that is before we had the 2 flat tires, about 5 miles apart! When we got done fixing the second tire Wayne asked if we left anything behind? "Only our pride," I said laughing. Wayne left my place at 5 PM so I guess he got back into his old routine again! I had to say good bye for now this morning to David Maki he felt he could get back into his cabin in northern Minnesota on the Vermillion River. I sure hate good bye's I sure appreciate David for his friendship and all he has done for me. After that I had to get planting barley on the organic farm. Joe had worked up 40 acres the day before while Wayne and I went down to Albert Lee, and God willing we will get that planted today! The Lord was willing we got the 40 acres planted even though a couple of time I almost stopped because of rain. It never did rain hard enough to stop planting. I am going to eat one meal late this evening and one early in the morning than continue on with my 22 fast which will be my last time. I will finish Saturday morning sometime April 16. I never planned it this way, but It will be 2 months and 2 days since I started this fasting.

I started the last 48 hour time of this fast on Thursday April 14 at 6:30 AM that would be Yesterday morning. I was going to wait until 7 o'clock, but something was telling me I better make it 6:30! So whatever is up with that? Once again I am extremely tired and called Jeff and Tom Kennebeck for prayers, that I finish this fast. We didn't do much yesterday because of misty rain. I did move some large stones and chunks of cement down by the old barn foundation. Today I followed Joe to have his transmission in his car looked at. I passed my neighbor Mr. X again as he was coming out of a bank why do I keep running into this person? I sent Joe out to do some field work at the organic farm while I took a part from the grain drill to my neighbor David Gibson's to be welded. While David fix I was listing to the radio the announcer was saying that it was the 99 TH year anniversary of the sinking of the Titanic. I was asking my neighbor David when he thought our Titanic would sink? We both just laughed, but I'm sort of thinking it may be sinking already! Later that day Jeff also mentioned that it was the 150th anniversary of the end of the civil war, so I thought that was interesting! We got Just 3 rounds planted on the field that Joe had worked up and got rained out, and as I write this the rain is turning to snow, and the ground is getting white! I have about 10 hour left and I am done fasting. The only word I can think of is Hallelujah! This morning Saturday April 16th I woke up early it is 3:37 am just under 3 hours of fasting yet to go. It's interesting the sense of urgency has been replaced with an overwhelming sense of accomplishment I feel more confident than ever that God has allowed great things to happen in the spiritual realm, because of this fast not because I did anything but because he did! I look forward to the future of things no matter what goes on!

> ***Psalm 65:1-5:*** *Praise is awaiting You, O God, in Zion; And to You the vow shall be performed.* **2** *O You who hear prayer, To You all flesh will come.* **3** *Iniquities prevail against me; As for our transgressions, You will provide atonement for them.* **4** *Blessed is the man You choose, And cause to approach You, That he may dwell in Your courts. We shall be satisfied with the goodness of Your house, Of Your holy temple.* **5** *By awesome deeds in righteousness You will answer us, O God of our salvation, You who are the confidence of all the ends of the earth, And of the far-off seas;*

Gods peace and abundant blessings to you all!

Your brother in Christ,

George Denn

Psalm 73: *1 Truly God is good to Israel, To such as are pure in heart. 2 But as for me, my feet had almost stumbled; My steps had nearly slipped. 3 For I was envious of the boastful, When I saw the prosperity of the wicked. 4 For there are no pangs in their death, But their strength is firm. 5 They are not in trouble as other men, Nor are they plagued like other men. 6 Therefore pride serves as their necklace; Violence covers them like a garment. 7 Their eyes bulge with abundance; They have more than heart could wish. 8 They scoff and speak wickedly concerning oppression; They speak loftily. 9 They set their mouth against the heavens, And their tongue walks through the earth. 10 Therefore his people return here, And waters of a full cup are drained by them. 11 And they say, «How does God know? And is there knowledge in the Most High?» 12 Behold, these are the ungodly, Who are always at ease; They increase in riches. 13 Surely I have cleansed my heart in vain, And washed my hands in innocence. 14 For all day long I have been plagued, And chastened every morning. 15 If I had said, «I will speak thus,» Behold, I would have been untrue to the generation of Your children. 16 When I thought how to understand this, It was too painful for me—17 Until I went into the sanctuary of God; Then I understood their end. 18 Surely You set them in slippery places; You cast them down to destruction. 19 Oh, how they are brought to desolation, as in a moment! They are utterly consumed with terrors. 20 As a dream when one awakes, So, Lord, when You awake, You shall despise their image. 21 Thus my heart was grieved, And I was vexed in my mind. 22 I was so foolish and ignorant; I was like a beast before You. 23 Nevertheless I am continually with You; You hold me by my right hand. 24 You will guide me with Your counsel, And afterward receive me to glory. 25 Whom have I in heaven but You? And there is none upon earth that I desire besides You. 26 My flesh and my heart fail; But God is the strength* *of my heart and my portion forever. 27 For indeed, those who are far from You shall perish; You have destroyed all those who desert You for harlotry. 28 But it is good for me to draw near to God; I have put my trust in the Lord God, That I may declare all Your works.*

As I sit here this winter day, I just want to sit here and give thanks to God! I haven't written anything since April I was waiting to see how my year played out. Nothing went anyway how I would have imagined it would by any stretch of my imagination. After cutting firewood for most of April and into May, on the evening of May 16 I had to sit here and watch and listen as others worked the fields around my house instead of me. No more would I work the friendly fields I had known for 49 years. I think if someone would have just shot me I would have hurt a lot less! I didn't have the heart to cut anymore wood. It was time to start planting pumpkins anyway, and the more that I could be away from here, the better it was for me. No longer did I feel at piece here, the place where I had witnessed it for so many years. It is interesting how quickly a person's life can change! Several days later my hired man Joe came and told me he was moving back to his parents place to help them get their farm established. I told Joe that what he was doing was very admirable as there was nothing better than to honor your parents. I could see it then that this would be the last year Joe would be working for me, the whole episode just saddened me greatly! I guess that's the only way I can explain it .Not since the days of my conversion did I wonder about God? What was he up to who is he really I had followed Jesus to this point, now he seems to be nowhere, and cares little how I am hurting! For more than a week I would ask myself who is this God I believe in. Is he some sort of spirit that just shows up when he wants to do something in your life and leaves? Just who is God anyway? Obviously he is not what I think he is. I told God one day that he could go and get himself another puppet, as I was tired of him playing with me. One of those days as I was feeling these things, my friend Gordy called me. He was responding to an e-mail I had sent out. Gordy said I thought we had established that everything in our lives was held together for Christ and by Christ? He was referring to Colossians 1 15-18:

> *15 He is the image of the invisible God, the firstborn over all creation. 16 For by Him all things were created that are in heaven and that are on earth, visible and invisible, whether thrones or dominions or principalities or powers. All things were created through Him and for Him. 17 And He is before all things, and in Him all things consist. 18 And He is the head of the body, the church, who is the beginning, the firstborn from the dead, that in all things He may have the preeminence.*

I told Gordy that I had but, living that out was quite another matter! It was then that Gordy revealed to me that his longtime friend «Moose» was dying. I knew Gordy was also going through a time of loss. Gordy was going down to see Moose before he died. I Told Gordy to tell Moose that I sure did appreciate knowing him, and that we had a lot of great times, and that I would see him on the other side of things, man what a sad week! Through all this time I never quit reading the bible. I guess in the last 17 years I have come to make it a habit. If I wouldn't have made it a habit I may have quit reading it at this time. One day I been reading in John 11 where Jesus had raised Lasuras from the dead. I had to admit that I believed that story to be true with all my heart, but I just couldn't rap my mind around it at this time. I began to realize that I probably felt like the disciples did when Jesus told them unless they ate his flesh and drank his blood they would have no part in him.

> ***John 6;53-64:*** *53 Then Jesus said to them, «Most assuredly, I say to you, unless you eat the flesh of the Son of Man and drink His blood, you have no life in you. 54 Whoever eats My flesh and drinks My blood has eternal life, and I will raise him up at the last day. 55 For My flesh is food indeed, and My blood is drink indeed. 56 He who eats My flesh and drinks My blood abides in Me, and I in him. 57 As the living Father sent Me, and I live because of the Father, so he who feeds on Me will live because of Me. 58 This is the bread which came*

down from heaven—not as your fathers ate the manna, and are dead. He who eats this bread will live forever.» 59 These things He said in the synagogue as He taught in Capernaum. 60 Therefore many of His disciples, when they heard this, said, «This is a hard saying; who can understand it?» 61 When Jesus knew in Himself that His disciples complained about this, He said to them, «Does this offend you? 62 What then if you should see the Son of Man ascend where He was before? 63 It is the Spirit who gives life; the flesh profits nothing. The words that I speak to you are spirit, and they are life. 64 But there are some of you who do not believe.» For Jesus knew from the beginning who they were who did not believe, and who would betray Him.

And like Peter I too had to come to the conclusion that Jesus Christ is the only one worth following. My identity is in him not this farm. So regardless of my circumstances and how I felt I knew I had to look to Jesus at this time for my answers as the world and other people had none! However my faith at this time in Father Son and Holy Spirit were shaken right down to my toes!

This spring was very late because it rained all the time. We finished planting pumpkins around the 9th of June. Joe had been sick for most of pumpkin planting so my friend Jeff helped get some of them planted.

As usual my finances were an issue. After 9 years of struggling under this issue, I made up my mind that if I didn't come out from under this issue, I was going to sell my equipment this fall after pumpkin harvest, and that would be that! Another issue that went on this spring was a vehicle issue. Joe's car broke down on him earlier in the year. I told Joe that instead of buying a car before he went on his trip to Italy it didn't seem to make much sense as he would use the money he would need for the trip to purchase a car. I had 3 pickups at the time I let him use the gold 1983 ford pickup, but we had to put about $600 in it to get it running again. Joe paid half and I paid half problem solved right! Wrong! I had bought a short box 1993 ford pickup from Joe so he could buy this car. I was driving that until the fuel pump went out and they wanted over $900 to fix it. I decided to junk that one out, and use the proceeds to buy gas to run the farm with! That left me with my old 89 ford pickup that was smashed and had no lights, a real wood hauler, but hardly road worthy. Joe was worried that I would want to take back the pickup he was driving. I was telling Joe that I wasn't going to do that, because I had people treat me like that when I needed help, and believe me that kind of help is like no help at all! I told Joe that I would clunk around with the old one until something came up for me. I am reminded at this time of the saying "no good deed goes unpunished". I just had to make sure that I got home before dark which was sort of tricky a few times! Northern light camp was just about here and I needed something better than what I had to make the 2.5 hour drive down there. I was looking on Craig's list early in the morning of July 3rd. Right away I found a 1990 ford pickup for 1000 dollars, and a yellow 78 3/4 ton ford. I called Terry to see if he could help me out with a loan. He thought he could so I let him do the dealing. because I had pumpkins to cultivate. On the 3rd Day of July I was able to buy a couple of pickup trucks again with the help of Terry and his mason jar money. I realized after we had done this it had been just two years earlier on the same day that we had bought 2 pickups. I scraped out the old junker I was driving, that helped us to buy gas for the week weeding pumpkins and also got me to Northern light camp. When I got there I found out staff was supposed to pay $150 to help with costs. I told pastor Troy I didn't have 150 dollars so the best I could do was for him to take the first $150 out of the pumpkin stand for this expense when we started to sell pumpkins this fall. My duties this year at camp were being a mentor so I didn't have the responsibilities that I had in past years which was kind of nice for a change. I didn't always agree with the direction our dorm was being lead , but I figured it was the counselors business not mine and he was the one that would be held accountable so I kept my thoughts to myself, and prayed about it. This year we had the Christian rock band Silver line as guests, and for 3 days they lead our worship service. I personally am not a fan of rock music, however not only did I tolerate it I actually enjoyed it, so to me that had to be a God thing! On the second night Ryan the lead singer gave a powerful testimony of how he and his wife had been experiencing God working in their lives the past couple of years. God was really speaking to me through Ryan. I thought it rather amusing that God would use a rock singer to minister to a farmer. Another night the whole camp was watching the Disney movie Tangled. Something happened to the audio visual equipment just about the time I was starting to enjoy the movie, so that was that! I have had the opportunity to be a part of the Northern light camp since it began. This year I see some signs of maturity for the camp. One of the staff this year had previously been a camper, and we had 3 baptisms that took place in the Root river. As usual camp was a great experience only in some ways different this year for me. When I returned home I had less than 2 weeks before heartland camp would take place Joe was still weeding pumpkins, but he too would be leaving for Italy for a month. The weather was unbearably hot, but as far behind as the crops were they needed the heat. Of course nothing ever wants to run too smooth between camps. Before Joe got the weeding done he got hives and was unable to work so I had to do his and mine both. I had to finish flame weeding, pickup wheat bundles and cultivating. As usual I was out of money and out of gas! I was asking God for some sort of provision for money or gas. I remember talking with Wayne Schwartz that morning he was wondering how I was doing I told him I wasn't doing to much as I was out of gas. Shortly after I hung up the phone Wayne called me back telling me he could spare me 20 gallons of gas until I could pay him back. God answered my prayer for Gas for the day. Jeff and Ernie helped me pick up the wheat bundles and the next day I cultivated pumpkins until dark. Saturday came and I had to seed some red clover and pick up some wire and fence post's where I had the wheat bundles Terry helped me with this I had just enough gas in the pickup to get over there and back. Terry was asking me what else I was going to do that day I told him I had 6 acres to cultivate but was probably close to being out of gas in the tractor the lawn mower has gas in it so I guess I will mow the lawn. Terry said why don't you just ask me for help, I felt God was asking me not to ask for his help as he would supply in another way. After Terry had left while I was mowing the lawn I thought maybe if I tipped the gas barrel I might find a little gas yet. I got 3/4 of a gallon but noticed the jugs that

Wayne had let me use one had slightly over a gallon of gas in it. Together that was 2 gallons I put this in the pickup and the gas gage read over 1/4 of a tank. Plenty to go to where the cultivator was and back again. When I got to the tractor I decided that I would not look into the tank I would just work until I ran out of gas or finished cultivating whichever came first! I had less than 2 acres left to go but for some reason the vines were bigger on this side of the field, and I felt that I was ripping out too many vines so I decided to quit. I pulled to the edge of the field and shut the tractor off. The big question was how much gas did I have left in the tank? There wouldn't have been enough to finish the 2 acres. Once again God was faithful for my needs and the pumpkins were laid buy until picking time! The next Morning I had to check the wheat field at Blase's. Wayne Schwartz and I made an agreement that he would combine my small grain while I was off to camp if I would return the hours he worked for me this fall with my guys and cut firewood for him after we got done picking pumpkins. The wheat looked ready to cut so I called Wayne and told him. I was talking to Blase when my dad called me. Dad was saying that my Uncle Lowell had this tree that he wanted cut down and that he would pay me if I would come and cut it down. I was telling Dad that I didn't have gas to get there. Dad bought me some gas and gave me $20 so I could do this work for my uncle. I worked for 3 hours on this tree it was extremely hot out I told my uncle that I wouldn't be able to finish it that day and I had to be in Illinois the next couple of weekends for camp, but I told my uncle I would be back to finish the job before I started picking pumpkins. Lowell paid me 70 dollars that day and Chris Schenk paid me for some hay the very next morning so I was able to pay Joe before he left that day to go to Italy. Boy looking back at it makes me think that those who think living one day at a time is something quaint, should try it sometime it might give them another perspective! I had to leave for camp on Thursday, but by the Sunday before I could see that unless I could come up with $1000 to pay up my bills for July, and buy Gas for the trip and back, I either had to ask for help or stay home. I told my friend Rick my situation within hours he called another friend Kevin they both agreed to loan me the money to be paid back at pumpkin time. They both sent their checks on Monday morning I hoped they would come by Thursday morning as generally it takes 4 days by mail when I get stuff from these guys! On Wednesday morning I had to take a hay rake over to Doris's farm so Wayne would have it if he needed it. After hooking up to the rake I could see it could use a few teeth so I drove it over to the shed put the teeth in which took about 45 minutes. I felt like a cup of coffee before I ;left so I went in and got one, and down the road I went. When I got to the hi way as I pulled on the road the mail man pulled up to my mail box I stopped to get my mail the only 2 pieces of mail were from Kevin and Rick which I felt was pretty miraculous as I said before it usually takes 4 days for mail to get to me from Rick and Kevin lives 300 miles from Rick! So I Had enough money to get to camp on, but I wasn't there a half a hour when Dave Salander and myself were going down to the worship hall to set some stuff up. Dave said a matter of factly «George I hear your having some tuff times right now here take this and pay me back when you can, and shoved some money into my hand it was $500 dollars! Later on at camp Bernie Bryant was telling me that some girls wanted a copy of my books Bernie wanted one too, but I didn't have enough with me. I told Bernie that he would have to give me his address so I could send him a copy. When Bernie gave me his address he also gave me a $100 bill and told me just to put that in my pocket! God spoke to me at that time that my books were unique and if I would just give them away that he would send me money from other sources! I have been doing that ever since. It sure was nice being at camp I felt like I had died and gone to heaven absolutely no stress level here! One day I was taking campers and staff to paintball. Ferris, Jordan and Britta were in the front with me I was asking Britta if she knew what the counselor's choice for the evening was. She said we are having a movie. I asked what was the name of the movie was. «Tangled,» she said. I thought how interesting that was the movie that I didn't get to finish 2 weeks ago at Northern light camp. Tangled is Walt Disney's 50th animated film. As I watched the movie God was speaking to me through it. All the experiences that had been happening to me back home I realized that somehow I was going to get through this year God was creating a way through my circumstances for me, and I will get to where ever he wants me to be where ever that is! It sure was hard to leave camp and go back home to what I had left behind. I don't think I cried so hard in all my life! Todd Woods was saying that they were giving away the food that hadn't been used at camp and that I should go and get some of it in case hard times lasted a while longer. I took a 50 pound bag of potatoes. I saw Stephan and Lilly and stopped to say good bye, Stephan asked if I needed any money. I told him that I didn't think so, but he too gave me more than $200. I didn't know it then but the money all those people gave me at camp was all the money that would come to me in the month of August other than $175 for a load of wood to pay Dylan and 40 dollars for finishing my Uncle's tree, and a 50 dollar gift

I returned home from camp on the 7th of August. David Maki came back a few days later he was going to work for me in August in place of Joe, and stay through the pumpkin season. We had straw to bale as Wayne had taken care of the harvesting, but it would be September before I would receive any payment from the grain. Dylan had called me before I had went to camp wanting to come up too help for a few days. I told him that I would be at camp for 10 days but I promised I would call him when I got home. I had to drive 11/2 hours one way just to pick Dylan up, and we had breakfast at McDonalds. Dylan was telling me on our way home that it was his decision to go to church, and go to camp. Dylan was telling me that he either had to go towards God or just become a druggie like his friends. I told Dylan that I thought he made the right decision. Dylan helped David and I bale straw and the 10 acre slough that I still rent here. I overheard Dylan tell David that he thought I was special. I just laughed to myself and told David later that I hoped Dylan didn't mean I was retarded or something! A guy by the name of Mike called for a load of fire wood so I was able to pay Dylan before he had to leave. Dylan left a little sooner than he originally planned. His parents kept calling him so I'm not 100% sure he was supposed to be here. I never had a check in my spirit about the situation so I will leave it at that. We were to meet Dylan's Mom in Waseca a town 30 miles west of me. Dylan was telling me that he would never forget this day. He was telling me that I made him rake the slough when he didn't want to. He said he wasn't doing a very good job and had to ask God to help him and he did! I was thankful that someone bought a load of wood so I could pay Dylan so I guess we both saw God at work this day! I had promised David Maki that I would have 3 months' work for him. I was figuring on remodeling my bathroom, as there had been nothing done in there for 30 years, and it was way past time! David was here for 2 weeks and I was unable to pay him! Chris had bought some more hay, and he was unable to pay me! And so the allusive money story goes on! Another instance that had taken place about this time was I had just been up to Gordy's fishing the morning I was leaving for home I got a call from Mary Porter. Mary was telling me that her grandson Connor had wished to see me before he went back to San Diego. I said that would be fine that I would be home around 1 PM. When I got home here Mary and Connor were here talking with David. We went to eat at the Happy Chef restaurant in North Mankato. Afterwards I took Connor by my pumpkin fields and was able to find him a couple of ripe pumpkins. Before Mary and Connor left for their 2 hour drive back to Mary's house I was able to pray for Connor. Also I gave Connor a copy of both of my books. I had pretty much forgotten about the incident until about a week later. We were putting up hay over on Doris's farm Chris asked me if I could go get him a couple of more hay racks. I told Chris I would but I was out of Gas in my pickup so he would have to give me $20 so I could make the trip home and back. Later on in the Day I had to send David Gibson back to get his duals for his baling tractor as it was a marsh, and he kept getting stuck. On our way home that night I noticed my gas was about empty again The 2 David's had used my pickup to go and get the duals. I kept praying for provision and was thinking of asking David Gibson for some gas, but God was impressing upon me that he had something better in mind. When I got home in the mail box was a letter from Mary Porter thanking me for the time spent with Connor ,and a check for $50 towards my next book being published! I knew Mary wouldn't mind if I borrowed the $50 for gas now! It sure was neat to see how God works through different people to get you through your circumstances! I remembered that tree at my Uncles that I promised to finish before pumpkin picking started. It had been more than a month since I had been there. David was telling me that morning how good a strawberry shake would taste to him, and the Lord answered his thoughts. We pulled into the Dairy Queen on one of our trips to the dump, and had a treat on my Uncle. When we finished My Uncle asked what he owed me. I told Lowell that he should give David $40, my dad $20 for gas and me $40. My $40 went directly into my gas tank! I was glad that David got something even if it was just $40! Joe came back to work on the 25th of August. David and I were already putting the signs on the hay racks for the pumpkin season. One day Joe, David, and myself .were taking a load of hay up to Scott's for his pumpkin stand in Lakeville. I was telling Joe and David that we should pray for some workers as we would not be able to pick 60 acres of pumpkins with just the 3 of us. About 20 minutes later I got a call from a guy by the name of Chris asking if I was going to need any help this fall. I was telling Chris that he was almost an answer to prayer as we hadn't prayed yet! I was confident that Chris's call was a God thing as Chris had worked for me 2 years earlier, and was a good worker. I wasn't expecting the pumpkin crop to be very good this year as the rest of the crops weren't. The Last weekend of August I took my Dad up to Gordy's fishing for his 80th birthday present from me. Dad had made the statement that it had been 30 years since he had went fishing. So I was glad to see that he caught some fish. It had been a special weekend for the both of us! I hadn't told anyone yet , but at that time I was just planning on getting through the harvest season and then I would sell my equipment and that would be that! September finally got here, and I got some money for the barley crop, and pumpkins were selling as well so I was finally able to pay my guy's. Poor David had to work for more than a month before he got paid! He told me that he had thought about quitting, but he didn't have enough money to get home on! Just as things started to look brighter on the 10th of September Minnesota put out a state wide freeze warning! A freeze at this time would finish me off! The squash were nowhere near ripe, and they can barely stand a frost let alone a freeze. I called everyone I could think of telling them to pray for my situation. Pumpkins can stand temps down to 27 degrees, but if it got much colder than that they would be shot! Well, all we could do was wait and see what happened. My guys and I could do no more than cover the stands. As I was making my way to pull tarps off the stands in the morning I drove by the pumpkin fields I could see that the leaves were froze and 90% of the pumpkins were still green! I was on my way to Wells Fargo bank to cash a check for some hay that a neighbor had purchased. Personally I don't like Wells Fargo banks every time I go in their bank to cash a check they want me to open an account. this time was no exception! I was ushered over to a guy by the name of Art. After Art asked me a lot of questions, I thought being I was in a hurry today, I would just make life a whole lot easier for myself and let Art open an account for me! In this process we found out that we both were Christians. Art is originally from Oklahoma. I was telling Art that I had a friend Brad that is from the Grove and Vinita area, he knew where that was. Before I left the bank with my cash and new account, Art prayed for my business I figured at this point I could use all the help that the Almighty wanted to send me! As Art was praying my cell phone was ringing in my pocket. I silenced the phone and let Art continue his prayer. After I left Art and got to my pickup I found out that a lady from the channel 12 news was trying to get a hold of me so she could do an interview. I told her that I could meet her in a few minutes at one of my stands in the Church of Christ parking lot in Eagle Lake. I also told her that I may look a little wild as I didn't have time to get ready for the interview. That's the way my life is any way what you see is what you get! By now a few people had phoned me to see the extent of the frost damage. I had told them that the pumpkins had froze but it would probably take a few days to see what the extent of the damage was in them and I hadn't been over to look at the squash yet, but I wasn't expecting much, because everywhere you looked you could see that it had frozen! I told the lady in the interview that it looked like it would be the second poorest pumpkin year that I have had in 12 years. I also told people that they should get their pumpkins bought early if they wanted to have one this year That's how bad things looked to me at the time of this interview. Now with that behind me I could finally drive to my field near Elysian to see how my squash turned out! When I got over there I couldn't believe my eyes. Not a trace of frost anywhere on the whole field! All the leaves were green and erect just as they should be, and yet it froze no more than a 1/2 mile in either direction from the field! I called Dave Holmes in Illinois to spread the news that we had a miracle God had answered our prayers! I also called Doug Johannsen from Anoka Minnesota to spread the word to others that God had answered our prayers! Somewhere about this time a guy by the name of Don left me note that he would be interested in purchasing 40 acres of standing red clover on my organic farm. Don and I met the next morning at the field to look at it. I was telling Don that I would need $100 per acre we both figured he would get enough hay to come out at that. Don told me to stop in and he would give me a check. I did, but he told me to hold it a few days until he could transfer some funds which I did. About mid-week Blase called in need of some rent money. I told Blase that I should be able to pay him shortly, but just not that day. After I thought about it I called Blase back and told him that I found a way to pay him. I would just sign the check and give it to Blase I figured that Don should have made the check good by this time. So that's what I did. In the meantime Don cut the 40 acres and baled the hay that was on it. He got a whole 23 bales far less than what we both figured! Don stopped one day and told me that he would just give me the hay if I would give him his money back. I told Don that I couldn't do that, because I had signed the check and given it to Blase. Don said that he had never made the check good, but he supposed that the bank would cover it, but that would be an extra charge! At this time I just laughed, and said "oh what a tangled web we weave!" I assured Don that I would treat him right I just needed some time to think what to do, because right at the moment I didn't have $4000 on hand. This is what I call a how would Jesus handle this situation! It was something I pondered over for a couple of weeks all along looking to God for the answer. An answer did come. I offered Don some year old hay for $100 a ton I would haul it also for that price and pay him custom work for what he did over on the 40 acres completely resolving him of the earlier deal. At first Don just wanted his money back, but I pointed out to him that it was he that wanted the hay and sought me out in the first place and also that he would be helping me out if he agreed to this deal instead by taking the year old hay off my hands. Don finally saw things my way and excepted my terms. The balance of the year old hay was sold about a week later for 100 dollars a ton. What a time trying to sell

Hay! Trying to make money is like pulling teeth! I was debating whether or not to hold the 8th annual pumpkin thing this year. I thought it would be somewhat harder as everything was away from here this year, but the money that is raised helps to lower tuition at our winter Snow Blast camp. I decided we would have it as it would probably be the last time we could if I can't find some land to rotate the pumpkins on. I was looking forward to having Ky come this year. I had called him 2 times and he hadn't returned my call, but his mom Denise called one day and said they were both going to come. As usual Tom and Sandy drove down from Orr Minnesota. They brought their grandson Ezra along with them. I was wondering how much we would get picked as we started Saturday with only 5 people, and the three guys that worked for me. As the day wore on, more people showed up. Troy Peterson came as well as Todd Fox and his 2 daughters. Mike and Daniel Haack also showed up close to noon. So at lunch time there was quite a crowd. Todd Fox's church sprang for lunch and we had submarine sandwiches, chips and drinks. It was fun to hear some of the guys tell stories of things that had happened when they worked here. Troy and Ky were reminiscing. As the afternoon wore on we got a lot of gourds, squash, and pumpkins picked. I was mindful of the time as Ky and Denise had to leave for Wisconsin around 5 PM. The whole day seemed rather miraculous. We started with 8 people, and ended up with a crowd! Dinner was taken care of, and to have Ky back for a visit was a special day for me, and a lot to take in. I said so long to Ky and Denise I would see them next weekend down at the Wisconsin Dells, Ky and I would be sharing a room. The next day we had a bunch of folks show up from Doug Johannsen's 2 churches up in the twin cities, and Troy and Ian from Spring Valley. It seemed that we had a whole lot of people show up, but couldn't seem to get organized too quickly. When it came time to eat we used Doug's tail gate on his pickup for a table. One person had bought chicken, another some rolls, and another some veggie's there was sodas and water to drink. I asked the blessing on the food it didn't look like there would be enough to go around, but in the end everyone had plenty it reminded me of when Jesus fed the 5000 people with 5 loves and 2 fish! I could see God's hand on the day different from the previous day, but all as rewarding! On Thursday morning I was off to the Kalahari hotel in the Wisconsin Dells where my church holds its fall celebration. I saw Ky right before church services, I was talking to Theodore. I asked Ky if he remembered Theodore. Theodore and I worked together at Snow blast back in February 2005. Theodore was one of my co-counselors that year; Ky was one of the youth that was in our dorm. Ky said that he remembered him after I told him who he was. Theodore was telling me that he would be leaving in the morning, so I wanted to make sure I gave Theodore a copy of my 2 books, because who knows when I would see him again. To be funny I bought Rick and Kevin each $500 in $1 bills to pay them back for the money they had borrowed me last summer! I also bought Dave Salander $500 almost in $1 bills. Dave wasn't there so I waited until the last day and gave them to Dave Holmes to give to Dave Salander when he seen him. I had a great time sharing a room with Ky, it seemed like the talk we had was supposed to happen. I was telling Ky that if he had anything against me yet from the summer that he stayed at my place that I just asked for his forgiveness. All I had to go on at the time was I asked God every day for wisdom and knowledge how to handle him. I told Ky that he had took 10 years off my life the 8 months he was here. Everyone had him wrote off including the psychiatrists he was going to see at the time! Everyone that is, except his mother, and me. I got to know Ky from the youth camps I served at, Ky would come as a camper. As I got to know him I really didn't think there was all that much wrong with him except he could use a little attention. I asked Ky if he ever hugged a cactus, because that's how easy, he made my life that year! I also told him how after I had it, and blew up at him one evening in October I walked out of the house and looked to the sky and told God how sorry I was that I had failed! But God knew what Ky needed, and me blowing my stack was exactly what he needed! I also told Ky how we took the sugar away from him, and how quickly he had changed after that, but it took us those many months to find out that sugar was messing him up as it was working against his medications he was taking at the time! I was telling Ky that even though I had to go through all of that, he had been worth it! Ky said that he really needed to hear that. Ky thanked me for being so hard on him, we gave each other a hug and I said a prayer for Ky. When I think of Ky I will always be glad that God included me to be a part of his life, I always feel that these things are the deeper part of what goes on at camp that few people see! One night I asked Ky if I could take him out to supper, as I didn't have any other plans for the evening. While we were eating I asked Ky what he would think about becoming half partner with me in my pumpkin business. Ky said he would say no right away so I would not be disappointed later on! I just sort of laughed, and said «well you know me Ky I don't take no for an answer»! I told him that he should pray about it, because if God wasn't behind the reason he would come I wouldn't want him there anyway! I just felt God wanted me to ask him that. Generally when God wants us to do something the first thing we do is

Run in the wrong direction. I say this with myself in mind mostly! When I got home from the Dells My cousins from Illinois were here. On Sunday evening I took Donna, Joan, Nancy, Kyla and Dad on what I call the "run". I have to do this for 61 days collecting the money and looking to see what the stands may need. After we did this and later counted the money back at my Dad's place, let's just say that it was an experience for them that they won't soon forget! We were all glad to have the time spent that evening as we thought we were going to miss each other because of our schedules. Pumpkin season went on without a hitch that is until Oct 14. Chris had set up a fund raiser that afternoon picking pie pumpkins the proceeds would go to a friend of Chris's needing medical attention. They must have picked 7 loads of pie pumpkins. I paid Chris for the fundraiser, and that was the last we would see of him. Monday Chris didn't come to work. Tuesday I tried calling him, but couldn't get a hold of him. Wednesday I got a hold of his Mom. She said all she could tell me was that last night Chris stole $100 and a bottle of vodka from her. I guess Chris is done helping us! On Saturday David's back went out, that left him unable to pick. That just left Joe and I, and we were getting pretty worn down too! Jeff would come and help us pick and Scott's nephew Joe needed some work so we eventually got done picking! Joe and David went to work on the bathroom next, I finally had the money to do it! We tore everything out tub, toilet, sink, floor we re did the whole room! It was somewhere at this time Joe told me that he wouldn't be working for me next year. I told Joe that I had known that for some time. When you have had as many guys working for you as I have they start giving you signs whether they realize it or not. It was interesting I had read this scripture not much more than a week before that!

> ***Isaiah 16:14:*** *14 But now the Lord has spoken, saying, "Within three years, as the years of a hired man, the glory of Moab will be despised with all that great multitude, and the remnant will be very small and feeble."*

Joe Had worked for me for three years. I truly will miss him not being around here it just won't be the same!

I guess I have said that about all of the guys who worked for me over the years, so I wish Joe well, but it is hard not to be sad about it all. David Maki left us on the first Monday in November, but Ethan came back to help finish with the work that we still had to do. The first thing we had to do was Joe, Ethan, Jeff and myself had to go over to Wayne Schwartz's and cut wood for 3 Days. This was to pay him back for the time he spent harvesting my crops while I was off to summer camp. My body was tired, and about the last thing I wanted to do was cut fire wood after all that pumpkin picking! I was glad when we completed the task. Ethan stacked all the wood that needed it, and split the four loads we bought home from Wayne's. The last day Ethan was here we got into a conversation. I was telling him that of all the years of my farming days I have enjoyed the ones most that I got to work with all the young guys that have worked for me here. When I think of all the things that got accomplished most of the stuff I never would have gotten around to by myself! It really does amaze me. Looking back at it each one seems to have been sent at a special time to do a special task, but just as God gives he also takes away. Why does one seem harder to take than the other?

So for the big question how did I turn out this year when everything was said and done? Well the pumpkins did about $15000 better than last year, and I have all my expenses paid from last year.

I bought a newer pickup than I have had in years, and paid cash for it. I have the publishing paid for this last book. I have some cash in my pocket and $25000 worth of hay and wood to sell for next year. So I feel that there has been a paradigm switch that has taken place.

> ***Psalm 126:5-6:*** *5 Those who sow in tears Shall reap in joy. 6 He who continually goes forth weeping, Bearing seed for sowing, Shall doubtless come again with rejoicing, Bringing his sheaves with him.*

I am better off right now than any other time in my farming career, and all with 1/3 less land, perhaps there was some sort of curse tied to the home place? If there has been one lesson I have learned this year, you don't have anything unless God allows you to have it! At present even though the year has been good for me I am having a hard time forgiving the people that took the farm away from me I guess God will have to help me forgive, because it is beyond my power to do so. For some reason this has been the hardest story I have had to write, and it's the only story that I didn't want to write. Quite possibly it will be my last, for only God knows if there will be any more stories to write about. I never expected to write anything, let alone 3 books! I guess if I write no more these 3 books are my testimony of how our Triune God (Father, Jesus, and Holy Spirit) have worked in my life. These are my pillar of uncut stones!

Gods' peace and abundant blessings to you all!

Your brother in Christ,

George Denn

Epilogue February 12, 2012

A couple of weeks ago my friend Wayne called me telling me that he decided to rent out his farm. Wayne's farm is 80 acres of plow land. He was telling me that he was going to see if his neighbors that farmed organic would be interested in farming his place as Wayne was an organic farmer as well. I mentioned to Wayne that if his neighbor wasn't interested in renting his farm I would be interested as I could use the land to continue my pumpkin raising on. This I left up to God as I was sure his neighbor would jump at the chance as farmland doesn't come up this way for rent very often! Nowadays it usually goes to the highest bidder. Wayne's neighbor thanked him for thinking of him, but after a few days thinking of it he declined Wayne's offer thinking it may disrupt what he already was farming. Wayne called me and told me this, and asked if I wouldn't mind if he talked with his brother who farms with chemicals. I told Wayne with a chuckle that it wouldn't bother me, as I hadn't had my heart set on it yet. I just left it in God's hands I figured if he wanted me to farm over at Wayne's he would make it happen instead of me trying to make it happen! I was wondering that evening how I would rent Wayne's any way as my hay hadn't sold anyway, and I was getting mighty tired of the scenario of not having the funds I need for springtime, so tired that I told God that morning about my frustration of it all. That evening a man named Zach called from Goodhue MN wondering if I had any organic hay left. I told him what I had for sale and the price. He told me that he would take it all! Something I felt was from God as I hadn't sold any hay in a while. Wayne had asked if I would stop over the next day and we could talk about renting his farm. After lunch, Wayne and I walked over some of his land I sure could feel Gods peace at this time I even mentioned to Wayne how it sure felt peaceful here! Before we started talking we said a prayer asking God to enter into our conversation I gave Wayne my proposal something I felt God was directing me to say. Before I finished talking to Wayne I told him I didn't want to rent his farm unless he felt peace about it all and directed his attention to Luke 1:78-79:

> *78 through the tender mercy of our God, with which the Dayspring from on high has visited us; 79 to give light to those who sit in darkness and the shadow of death, To guide our feet into the way of peace."*

I find that interesting, because that too was given to us by a Zach! Later that evening, Wayne called and asked if I could pray for him as he was full of anxiety. I prayed for Wayne and again directed him to a scripture in Philippians 4:4-7:

> *4 Rejoice in the Lord always. Again I will say, rejoice! 5 Let your gentleness be known to all men. The Lord is at hand. 6 Be anxious for nothing, but in everything by prayer and supplication, with thanksgiving, let your requests be made known to God; 7 and the peace of God, which surpasses all understanding, will guard your hearts and minds through Christ Jesus.*

I was telling Wayne that it seemed that God was telling us through that scripture not to settle for the anxious feelings that so often we feel when he has something way better for us instead!

Just yesterday morning Wayne called me and asked if he could stop in as he was feeling very at peace with renting his farm to me. Wayne and I talked about the particulars of renting his farm, and in the end we prayed about it and shook hands over our deal that's all Wayne and I felt was necessary! I felt this arrangement with Wayne felt more personable than any contract I had ever signed! I also realized at that time it had just been a year previous on February 14 that I found out that I had lost my home farm, and I had just witnessed God replacing that with a much better farm! All of a sudden I almost feel I could care less about this old place, and see God sending me more out into his world! As I finish with this third book I found this last story extremely hard to write. It took more than 2 months to complete. I feel the gift that God had given me after 12 years was starting to wane. I really don't expect to write any more stories, but this I will leave up to God. I never expected to write anything let alone 3 books! I know sometimes when I am down or discouraged I can re-read the stories I have wrote, and remember all that God has worked out in my life. I find you can see God (Father, Jesus, and Holy Spirit) in the times you are up like when you just witness a miracle, but also in sad times, like the loss of something, or someone very close to you. Often when he is pruning you it takes a little longer to see him until sometime has past! One of the things I have learned while I have been writing that a story may only be true at the time it is written. Time seems to change things, places are sold and have new names, people die, and even we change. I know I perceive God far differently now than when I first started to write! Time and circumstances have caused me to lean more on him and less on me! One day not so long ago Jeff Peterson, David Maki, and I were in my kitchen enjoying a cup of coffee, and the heat from the wood stove. I forget just what we were talking about, but I made this statement. "I am the least likely person to help myself out in this situation." After making that statement I realized I had said something profound, and Give the Holy Spirit the credit, because God really is the only one that can help us in a time of need! I hope you have enjoyed the story's if you are lead to comment please do!

George Denn
59381- 243 St
Kasota, MN 56050

CPSIA information can be obtained
at www.ICGtesting.com
Printed in the USA
BVHW030004040921
615742BV00001B/3